# 50 Green Projects for the Evil Genius

## Evil Genius Series

*Bike, Scooter, and Chopper Projects for the Evil Genius*

*Bionics for the Evil Genius: 25 Build-It-Yourself Projects*

*Electronic Circuits for the Evil Genius: 57 Lessons with Projects*

*Electronic Gadgets for the Evil Genius: 28 Build-It-Yourself Projects*

*Electronic Games for the Evil Genius*

*Electronic Sensors for the Evil Genius: 54 Electrifying Projects*

*50 Awesome Auto Projects for the Evil Genius*

*50 Green Projects for the Evil Genius*

*50 Model Rocket Projects for the Evil Genius*

*51 High-Tech Practical Jokes for the Evil Genius*

*Fuel Cell Projects for the Evil Genius*

*Mechatronics for the Evil Genius: 25 Build-It-Yourself Projects*

*MORE Electronic Gadgets for the Evil Genius: 40 NEW Build-It-Yourself Projects*

*101 Outer Space Projects for the Evil Genius*

*101 Spy Gadgets for the Evil Genius*

*123 PIC® Microcontroller Experiments for the Evil Genius*

*123 Robotics Experiments for the Evil Genius*

*PC Mods for the Evil Genius*

*Programming Video Games for the Evil Genius*

*Solar Energy Projects for the Evil Genius*

*Telephone Projects for the Evil Genius*

*22 Radio and Receiver Projects for the Evil Genius*

*25 Home Automation Projects for the Evil Genius*

# 50 Green Projects for the Evil Genius

Jamil Shariff

New York Chicago San Francisco Lisbon London Madrid
Mexico City Milan New Delhi San Juan Seoul
Singapore Sydney Toronto

**Library of Congress Cataloging-in-Publication Data**

Shariff, Jamil.

50 green projects for the evil genius / Jamil Shariff.

p. cm.

Includes index.

ISBN 978-0-07-154959-2 (alk. paper)

1. Green movement. 2. Green technology. 3. Energy conservation. 4. Sustainable living. I. Title. II. Title: Fifty green projects for the evil genius.

GE195.S47 2009

640—dc22 2008049640

McGraw-Hill books are available at special quantity discounts to use as premiums and sales promotions, or for use in corporate training programs. To contact a representative please visit the Contact Us pages at www.mhprofessional.com.

**50 Green Projects for the Evil Genius**



1 2 3 4 5 6 7 8 9 0 QPD/QPD 0 1 3 2 1 0 9 8

ISBN 978-0-07-154959-2
MHID 0-07-154959-5

**Sponsoring Editor**
Judy Bass

**Acquisitions Coordinator**
Michael Mulcahy

**Editing Supervisor**
David E. Fogarty

**Project Manager**
Andy Baxter

**Copy Editor**
David Burin

**Proofreader**
Grahame Jones

**Indexer**
Burd Associates

**Production Supervisor**
Pamela A. Pelton

**Composition**
Keyword Group Ltd.

**Art Director, Cover**
Jeff Weeks

The pages within this book were printed on acid-free paper containing 100% postconsumer fiber.

# About the Author

Jamil Shariff is a writer, educator, multidisciplinary freelancer and consultant. His passion is for environmental technology, especially the obscure but useful, like biogas digesters, sandbag houses, or velomobiles. He has been educated in music, English, and politics, and received his Masters in energy and the environment in architecture.

Jamil has worked with well-known environmental groups, like the Sierra Club, and the more obscure, like the Boiled Frog Trading Cooperative, and currently serves on the Board of Directors of the Toxics Watch Society of Alberta. He delights in writing to disseminate his experiences with new technologies, and designing and delivering technology-based courses.

His chronicled experiences range from sucking up used grease from restaurants to fuel the car he converted himself, to building parts of the fiber-glass weather protection on his velomobile himself. The courses he taught range from postgraduate classes on Stirling heat engines, to training sessions on biodiesel production for a local cooperative.

Jamil currently works with a sustainable innovation incubator in Ottawa, Ontario called the Boxfish Group, named for the angular yet rigid fish that car designers are mimicking. He works to bring deep technical understanding to policy conversations that can help government and businesses foster the innovations we need.

To my tenth grade English teacher, who wanted to fail me:

ttthhhbbbbbbbbbbbbbbbbbbbbth!

# Contents

# Foreword

People the world over are beginning to realize that a clean environment and a strong economy are two sides of the same coin. This realization, together with some ingenuity, will help to drive the next industrial revolution. We will see the creation of new environmentally friendly industries and buildings. We will set the foundations of a sustainable future.

Everyone can contribute to the global effort to reduce carbon emissions. By getting going we can make improvements in both our environment and our quality of life. New ideas and projects are emerging every day, including some of the technologies you will read about in this book. With more people reading and experimenting along these lines, we will see broader support for change at the individual and social levels. Many people want to make this change and start real progress on the road to a sustainable and prosperous society.

While I was Prime Minister, the Canadian government started the long road with Project Green. We need to make strides now down that road towards getting useful environmentally friendly technologies into the hands of all Canadians.

As Jamil mentions, neither politics nor technology alone will, in the end, be sufficient to meet the enormous challenges that our changing climate will pose.

Innovation and the commitment of millions of Canadians, the inventiveness of many so-called "evil geniuses" working away in labs and shops—that is what will make it possible to come out ahead and even more prosperous. But we do need to get serious about this now. Climate change is one of the biggest challenges humankind has ever faced. It challenges us to work together as a planet and as a nation as never before. But we need to make light work with millions of hands. We all need to get interested in seeing what can be achieved in our own households and businesses. Each of us needs to start somewhere getting on the road to a sustainable lifestyle, and there is something in this book for all. I hope you enjoy the book as much as I have.

**The Right Honourable Paul Martin PC**

Former Prime Minister of Canada

# Acknowledgments

There are always a great many people to thank for their contributions to any work of this nature. To begin with, I am indebted to Gavin Harper for thinking of me for this project and for putting my name forward.

I'm grateful to everyone at the Falls Brook Centre in New Brunswick, for building an interesting learning center and for their help in preparing parts of this book, specifically Terri, Jean Arnold, and Brent Crowhurst. I'd like to thank Dries Callebaut at WAW in Belgium, Bar De Wert of Aerorider in the Netherlands, Rod Miner at Lightfoot Cycles in Montana, and Krash for images of the velomobiles. I owe a debt to John Tetz for pictures and help with the steps involved in building with Zotefoam, and to Dave Eggleston at Pedal Yourself Healthy for help with sections on the Alleweder, as well as to Jurrian Bol at Dutch Speed Bicycles for images of the same. Gary Whitfield at Whispergen was very helpful in providing prompt information and really neat pictures and Stirlingengine.com was kind enough to send a model for my use.

Carl Georg Rasmussen is an inspiration to me for his years of dedication to building a desperately needed vehicle by hand, and I'd like to thank him for teaching me his methods and inspiring me, as well as providing timely photos for the book. Myles Kitagawa has also been an inspiration to me for years, for this as well as permission to quote him, I am grateful. Raphael Khoudry gave eager assistance in compiling images of some projects. Richard Hampton's forethought in keeping a tin-can engine around for years until needed is deeply admired, and his comments on early drafts were greatly appreciated. Vinay Gupta, as always, was ready to do what it took to help have the Hexayurt widely adopted, and Rob and Sky Bicevskis went out of their way to make sure images of the fire piston were good enough for print. I'm very grateful for all of their help.

As ever, my family has been great throughout this journey, and alternately supported me and left me alone when needed. Thanks to all of you, especially to Nashina who looked over some chapters to make sure her older brother wasn't saying silly things.

I'm grateful to Judy Bass, my awesome editor, and Andy Baxter at Keyword, for their expert assistance throughout the publishing process, and to David Fogarty, Pamela Pelton, and David Zielonka, and the other members of the team at McGraw-Hill that guided me and contributed to this work. Last, but certainly not least, I would like to acknowledge the work of Brian Guest, who made the awesome Foreword possible, and the rest of the Boxfish team who are ensuring that my life remains full of interesting technologies and projects.

I'm sure I've missed a couple very important people to thank, and I'm sorry I did, but greatly appreciate their assistance nonetheless. Of course, none of this would have been possible without hundreds of inventive souls who were experimenting and exploring issues and technologies that will be crucial to a sustainable future long before me, and thanks is not enough for their persistence.

# Chapter 1

# Introduction

## The environment

During the last couple of years, the environment has become a popular topic in the media. In particular, global climate change has been receiving increased attention, due to the mounting evidence that its impact will be global and could be severe—even catastrophic—if mitigating action is not taken quickly.

In the past, topics such as acid rain have received a lot of attention. This in turn spurred action to address the problem, and while the problem hasn't been solved, a lot has been done to address it over the years. The current media focus on climate change may lead to the same kind of increased action, helping to avoid some of its most severe impacts.

## Climate change

According to scientists, the warming of the climate, known as global climate change, is now unequivocal. It is increasingly clear that much of this warming is the result of human activity since the Industrial Revolution. Climate change results from the release of particular gases into the atmosphere, called "greenhouse gases" (often abbreviated GHGs); they act not unlike the glass on a greenhouse and trap the sun's heat on earth. These gases exist naturally as part of systems in the earth's cycle and scientists estimate that the earth would be about 30°C colder than it is today if these gases did not exist. The same scientists also note that human activity over the past 100 years has led to a perceptible rise in global temperatures. In 2008, the news is filled with stories about northern ice caps melting and glaciers receding, which are the result of the excess greenhouse gases that are emitted into the atmosphere as a result of human activity.

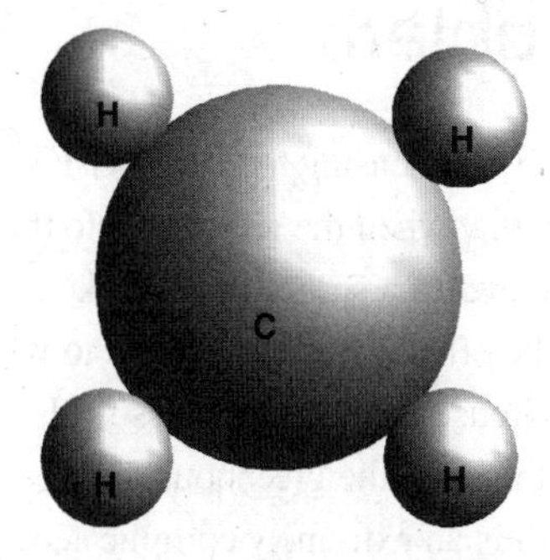


**Figure 1-1** *Molecular structure of methane ($CH_4$), a powerful greenhouse gas.*

The major greenhouse gases include methane (scientifically notated as $CH_4$—meaning one atom of carbon and four atoms of hydrogen bonded together in a single molecule of methane–see Figure 1-1), nitrogen oxides ($NO_x$—one nitrogen atom and one or more oxygen atoms), water vapor ($H_2O$—two hydrogen atoms and one oxygen atom) and carbon dioxide ($CO_2$—see Figure 1-2). Moreover, there are a number of entirely human-made greenhouse gases in the atmosphere, such as the *halocarbons* and other chlorine- and bromine-containing substances. These gases are emitted in varying amounts through all sorts of natural and human activities; the largest single human source is the

**Figure 1-2** *Molecular structure of carbon dioxide ($CO_2$).*

burning of fossil fuels. These are currently our prime sources of energy, and include coal, gasoline and diesel fuels, and natural gas. We use them to heat our homes, power our cars, and in giant power stations that provide the electricity needed to power our electronics.

## The problem

The problem with releasing too many GHGs into the atmosphere is that when the gas rises into the atmosphere it creates a sort of heat blanket, as we mentioned. The effect has been likened to what happens inside a greenhouse—where the glass walls trap the sun's heat—but in this case the greenhouse is the entire planet. Since the earth is an extremely complicated system, this doesn't necessarily mean warmer temperatures everywhere in the world, but that the earth's average temperature will rise, causing an increase in extreme weather events and potentially drastic changes to weather patterns such as droughts and hurricanes.

Recent scientific reports have shown that the projected impacts of climate change could be quite severe. Some expected changes include:

- shrinking Arctic and Antarctic sea ice
- more frequent heat waves
- more intense tropical cyclones (typhoons and hurricanes)
- plant and animal extinctions
- increased flooding due to sea level rise
- increased drought.

There are also sharp differences in how climate change will affect people across regions: those in the weakest economic position are often the most vulnerable to climate change and are frequently the most susceptible to climate-related damages.

## Upsetting the balance

The atmospheric balance is being upset by the extra amounts of carbon dioxide and other greenhouse gases being emitted into the atmosphere, largely as a result of human activity. Figure 1-3 shows, very generally, the

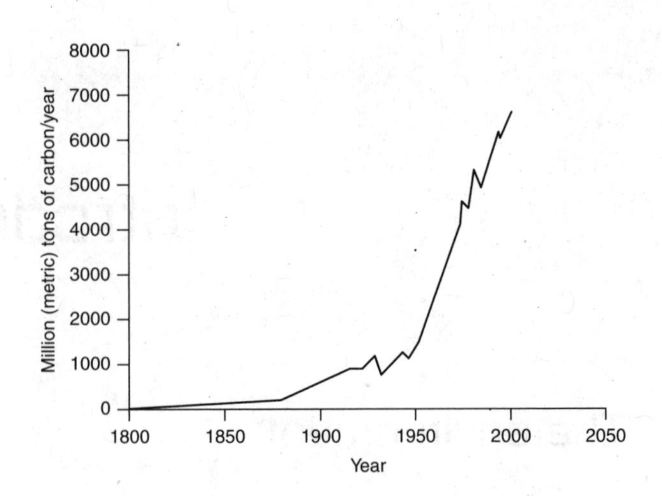


**Figure 1-3** *General trend in global emissions over the past century. Used with permission from Gavin Harper's* Solar Energy Projects for the Evil Genius, *McGraw-Hill, 2007.*

trend in our yearly emissions over the last century. The problem with the carbon dioxide that we are releasing into the atmosphere is that most of it comes from sources under the earth's crust (e.g., oil), where it has been stored for billions of years. Before we humans started digging into the ground looking for this black flammable material, most people used wood and other sources that came from recently living things. Trees absorb carbon while they are growing, and when they are burnt the carbon that they release is carbon that was recently in the atmosphere to begin with, so it is a relatively short cycle. On the other hand, when we burn fossil fuels we release carbon into the atmosphere that has been stored away by the earth for billions of years. This extra carbon is causing the earth's temperature to rise, resulting in serious environmental impacts.

So, what's to be done about it?

## Doing many things at once

There have been several international agreements related to climate change. The most famous of these is the Kyoto Protocol (named for the Japanese city in which it was signed), which committed all industrialized countries that signed on to collectively reduce their emissions below 1990 levels by the 2008–2012 period. While not all industrialized countries ratified the protocol, most notably the United States, and some countries that did ratify and commit to reductions are not on track to meet their commitments, the Protocol is considered by many

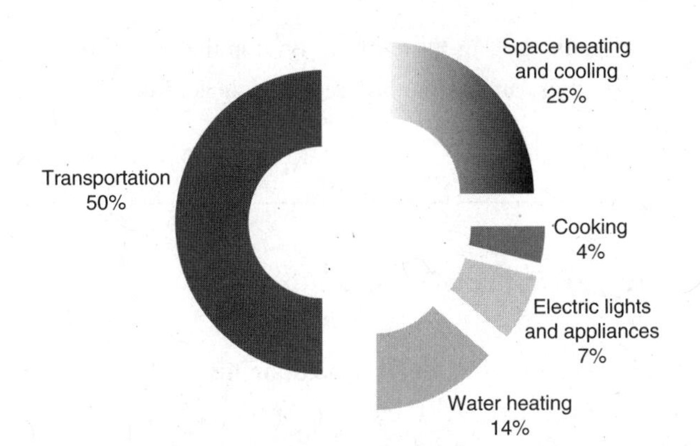


**Figure 1-4** *A general idea of the average North American's energy consumption pattern.*

to be an important first step towards avoiding dangerous climate change.

What else can be done, you might ask? Well, the truth is that each of us as individuals has been responsible for emitting several tons of GHGs in the atmosphere each and every year we've been alive. Figure 1-4 shows an approximate breakdown of how the average modern consumer uses energy. This gives us an idea of what areas of our life we should approach first, if we are serious about becoming green. We are each likely to feel the effects of changes to the climate, so it seems reasonable then that each of us should play a part in reducing and managing the changes to the climate. No sustainable change can take place unless we begin and end at the individual level.

There are a great many things that each of us as individuals can do, which can complement actions taken by our governments and the companies that support our lifestyle. We can start with small actions (Figure 1-5 is an example), to help us remember that we are concerned. These will be a start to becoming "light green," or a little green. We will discuss these simple options early on in the book, and then move on to looking at larger things we can do. Many simple actions, like installing energy-efficient showerheads or looking for local food producers, have groups that have formed around promoting these actions in their area. Readers could think about joining and supporting these groups—and if there isn't one operating where you live, perhaps starting one of your own.

**Figure 1-5** *Small actions to reduce energy are part of the solution, so long as they don't distract from the larger changes that need to be made.*

## From small to consequential

Small actions are just the beginning, and shouldn't distract us from the big picture. Getting enough environmentally friendly and efficient technology around us that will make a difference to the climate will require us to explore and install new technologies, like solar and wind energy. But, we also need to make good use of our existing technology and act on legislation that is relevant to technology, like feed-in tariffs or innovative bicycle policies.

The first section of the book concentrates on the smaller everyday actions that each of us can take to reduce our own emissions. We then move on to more in-depth projects, in hopes of providing the inductee into green living with an idea of where these little actions are leading. Throughout, we try to acknowledge that technology does not exist in a vacuum, and is therefore only as good as society knows how to put it to use.

Many of the countries that did sign up to the Kyoto Accord did so because they felt pressured by their citizens to react to what was commonly perceived as an urgent issue. An important way in which individuals can bring about change is through popular action designed to put political pressure on government representatives. In the United States, there are established methods of pressuring elected representatives, through elections or lobbying representatives. There are also other, less established ways to get attention, such as those pictured in Figure 1-6. Getting involved with local environmental activism is a good way to let your government know that this is an important issue to you, and that their time in power may be dependent on how much they are paying attention to this issue.

**Figure 1-6** *Sometimes the only way to get your voice heard is to use any means at your disposal, including taking it to the streets.*

> Never doubt that a small group of thoughtful, committed citizens can change the world. Indeed, it's the only thing that ever has.
>
> Margaret Mead

Corporations can also be pressured in this way, through campaigns that communicate the importance of this issue. Large and small companies have the ability to dramatically alter their behavior and choose less-polluting alternatives, but they often need incentives to do so. Sometimes this happens because of governments that set the regulations which determine how they operate, and other times by customers who can use their leverage (the threat to cease to buy products from a company) to effect changes. As the title of this section mentions, it is important to do many things at once, because it is unlikely that any one of these actions alone is going to meet the very severe challenges posed by a changing climate. Together they might have a chance.

## Lessons learned

The phenomenon of acid rain affected large parts of Canada and the United States, making headlines everywhere about 30 years ago. Acid rain is partly a result of the sulfur in the fuel used in coal-fired electricity-generating stations across the continent. Released as sulfur dioxide ($SO_2$) into the atmosphere when burnt, it is transformed into sulfuric acid ($H_2SO_4$) while traveling over long distances in the air, and is then deposited as acid rain over large areas. Acid in the rain reduced the ability of a variety of living organisms to grow, causing long-term effects on populations of fish and animals, and damaging trees and agricultural products.

### Online Resources

www.epa.gov/region1/eco/acidrain/history.html—the Environmental Protection Agency gives a good introduction to the problem of acid rain and some of the measures that have been useful in combating the problem.

Because the problem of acid rain crossed state boundaries in the United States, the federal government passed a series of regulations that have changed the way all coal-fired power plants in the country deal with their atmospheric emissions. Through international agreements—among the first international environmental agreements ever—both Canada and the United States agreed to pollution reduction targets through the mandatory introduction of new technologies. The agreement has prevented about 10 million tons of $SO_2$ from being released into the atmosphere by industrial electricity generators since it was signed. A common technology used to achieve this monumental environmental feat is called flue gas desulfurization (FGS), which is pretty neat. A wet scrubber type of FGS system uses a limestone slurry (think wet concrete), injected into a tall reaction tower where a fan sucks the heated gases from the combustion process. The sulfur dioxide reacts with the limestone slurry and becomes calcium sulfate ($CaSO_4$), which is no longer acidic and is useful for lots of chemical processes. Some power plants are now able to sell their calcium sulfate to other industries and recover some costs.

So a dangerous pollutant is removed from the chimneys of our power plants in a useful and profitable form. Nice to know that not everything that helps the environment has to cost us money. The other important thing to remember is that we had to know that sulfur was a problem in order to have something done about it. Most of us don't think very much about how our car's engine works, but that might change as fuel becomes more expensive and harder to come by. Or, maybe, it will be because we become more aware of what pollutants are coming out of our tailpipes. Other common acid rain causing emissions include nitrogen oxides ($NO_x$), which are emitted by, among other things, your cars. Because of similar legislation as that described above for power stations, most cars produced in the last decade came fitted with catalytic converters that use platinum and other metals to reduce the emissions that contribute to acid rain. Catalytic converters have been implicated in slightly lower fuel economy, so there is a tradeoff. However, the result is still that what is coming out of the tailpipe of an old car probably has more pollutants than a newer car, regardless of the fuel consumption.

What have we learned? It is not just about what type of engine or prime mover is used to turn one type of energy into another; it is important to be aware of the details. When burning wood in a fireplace, for instance, one might be tempted to feel good and green about using a renewable fuel. Many people will call using wood for heat carbon neutral, because the carbon being released by the wood was very recently taken out of the atmosphere (unlike the carbon contained in fossil fuels). But if wood is burnt in the absence of oxygen, a lot of that carbon-neutral carbon goes up the chimney as methane gas ($CH_4$) rather than carbon dioxide ($CO_2$), which results from a complete burn in the presence of oxygen. Let us focus for a minute on methane, because it is a pretty neat gas.

Depending on how much methane is in a mixture and where it came from, it can have a lot of names. Probably the one most people are familiar with is "natural gas," which is common in parts of North American cities as a home heating fuel, supplied through an underground pipe network. The natural gas supplied through these networks is almost universally a fossil fuel, which is found in the same wells as crude oil that is refined into gasoline. To some oilfield operators it was considered a waste product and vented directly into the atmosphere, until regulations stipulated that the gas should be flared—i.e., burnt like a torch. In parts of Alberta, Canada—where there are a large number of oil and gas deposits under development—it was possible to see flares scattered across the dark skies until recently. In many cases the gas is now captured and used, which is much better for the planet and better for us too, because we have another fuel source available. Contrary to what most people think, natural gas happens to be pretty clean in comparison with some of the other petroleum-based alternatives.

## The deal with methane

The problem with releasing methane gas into the atmosphere is that this particular gas has a really big impact on the greenhouse effect we discussed in the first section of this chapter. The Intergovernmental Panel on Climate Change, possibly the most influential group of scientists around the world, estimates that methane gas has over 20 times the greenhouse effect of carbon dioxide in the atmosphere. 20 times! That is really big, especially when you look at what methane and carbon dioxide are composed of (see Figures 1-1 and 1-2).

Note that in both methane and carbon dioxide, there is only one atom of carbon. Remember that methane has 20 times the greenhouse potential of carbon dioxide. Now think back to the wood fireplace you have at the cottage and try to remember what the dampers do: they reduce the air supply. When there is less air reaching your wood fire, more methane is produced. The carbon that the tree absorbed during its lifetime was probably carbon dioxide, but by turning the fire down for the night you could be re-releasing that carbon in a much more dangerous form back into the atmosphere.

## Technology and society

When companies and governments are pressured to change the amount of GHGs that they release for their activities, technology is often seen as the most expedient and relevant way to make those changes: e.g., switching to lower carbon-content fuels—from coal to natural gas (which contains less carbon and other pollutants for each unit of useful energy it delivers). The social aspects of using energy, for example driving less by choosing to use other modes (such as biking, walking or taking the bus) for some trips, can be equally important.

Failing to adequately heed both social and technological aspects of technology use can result in unintended consequences. Because we are trying to make fairly specific changes to our collective impact on the environment, it is necessary to be aware of both sides and not concentrate on either one too heavily.

For example, a more efficient car is only better for the environment if the person driving it continues to drive the same amount as they did before the fuel savings. Consider a person who was driving 100 km per week and spending $50 on fuel, who buys a new fuel-efficient car that allows them to travel 150 km on the same amount of fuel. The environment only benefits if that person continues to drive 100 km a week (saving money on fuel) rather than starting to drive 150 km a week because they can now afford it. Similarly, a more efficient furnace only benefits the environment if the occupants of the building don't use the fuel savings to raise the temperature during the winter.

## Starting now

So, while this book's focus is nominally on the technologies that can help you to reduce your emissions, it is important to realize that technology alone is not sufficient. Neither are legislation or social action likely to be enough on their own, but together, by doing many things at once, we might just start to make a significant impact. So, turn down the thermostat, put on a sweater, ride your bike back from the store, and then settle in to read the rest of this book by an energy-efficient light bulb. But if you don't have an energy-efficient light bulb handy, and your bike has a flat, don't panic. You would not be the first to have faced this problem, and it only takes a little evil genius to get going, so keep reading and we'll get there together.

# Chapter 2

# Transportation

## Getting about

Transportation accounts for about half of the energy consumed by the typical North American household. If we are serious about reducing our energy consumption and doing something about this climate change thing, we are going to have to do something about our transportation-related energy consumption. This is one of those things that is going to involve more than just a change in technology, like a hybrid or even hydrogen fuel cell. In addition, we also need to become smarter both about how we as individuals travel, and how we as a society accommodate different types of travel in general. Consideration of this large chunk of emissions is far too often shied away from by otherwise environmentally conscious people, because the scale of change required seems to boggle some minds. It need not be so.

Think about how you got this book. You may have traveled to a bookstore to buy it, perhaps you drove. Could you have walked instead? Why not? In too many cases the answer is because the distances are too great. This is particularly true in suburban and rural areas of the country. The widespread and growing use of the automobile for the past 50 or so years has closed what were once small but thriving rural and suburban commercial centers and increased the distance between locations people frequent, such as their home, office, and shopping centers. When there is not an alternative to a car available for people to use to reach those locations, a condition of automobile dependence is created. The alternative is to increase density and have those common locations we all visit spaced closer together (see Figure 2-1). Urban centers, by their nature, have a greater concentration of people and services, and people who live in such centers might have had a greater chance to buy this book from a store they walked to.

**Figure 2-1** *This scene from the busy metropolis of Istanbul shows what transportation options in an urban center look like: trains, cars, and pedestrians.*

## Green cities

Studies that looked at individual energy consumption patterns have found that the typical urban (living in a high-density city) dweller's energy consumption is far below his suburban or rural counterpart. The efficiencies in living in urban centers come from many areas, including transportation. Having many people on a train is far more efficient than having many cars on the road (see Figure 2-2), and having shops and houses close together means that people don't have to travel so far in the first place. This flies in the face of my early exposure to what being an environmentally conscious person meant. Like many Canadian children, our family spent the summers camping in the national and provincial parks. We lived in tents, with bugs, carried our own water to the campsite, went fishing, and enjoyed the great outdoors. When I started exploring the environment as an issue later on in life, I, like many

Figure 2-2 *Stuck in traffic? Think about whether you* really *needed to make the trip. Reducing our car travel is more important than improving the emissions of individual vehicles in order to reduce scenes like this.*

others, found a connection between my experiences in nature and this idea of an environment that should be protected. However, living in, playing in, and even seeing places where nature abounds, in the forest and the periphery of urban centers, can have particularly large impacts on our transportation-related energy consumption.

**Tip**

You don't have to hug a tree to be green: urban centers may not get all the credit they deserve for being environmentally friendly places to live. Benefits include lower transportation emissions, as well as lower infrastructure and service costs, and lower heating costs because heat radiates between buildings and apartments.

If the scientists are right, reducing the amount of carbon that we emit into the atmosphere should be the top priority for modern societies. If we do not manage to reduce the concentration of greenhouse gases in our atmosphere soon, we may one day face the prospect of runaway climate change. Scientists speculate that when a certain concentration of $CO_2$ in the atmosphere is reached, the rising heat of the planet will lead to more heat-trapping gases being released from natural sources. This creates a sort of feedback loop in the climate that kicks in after a certain tipping point. Such a feedback loop would decimate most of the life on the planet, along with the majority of us. So, while I enjoy the forests and the parks, I think twice about going to drive there, alone in my car, for the weekend. That is certainly part of being smart about transportation and a start on the road to becoming "green."

Anyone today could have ordered the book online and had it delivered, potentially saving a trip for you, but also possibly helping to create a more efficient distribution system. Retail stores aren't always the most efficient way of keeping a large warehouse of products in terms of energy consumption. Often, retail stores need to be located near to customers, with good transportation access, while warehouses can be located in more inconvenient locations. Sending goods directly from warehouses in regular shipments can at times be more efficient than sending goods to a store first to be collected. But that would have meant your knowing about it, or finding it before hand, rather than just stumbling across it on the shelves. I think this is an example of the social side to sustainability, which is the foundation of this whole "green" thing. Sustainable development proposes that we should give equal consideration to social, economic, and environmental aspects of everything (Figure 2-3). It tries to stand in contrast to everyday development, which is often concerned solely with the economic aspects and tends to miss the others.

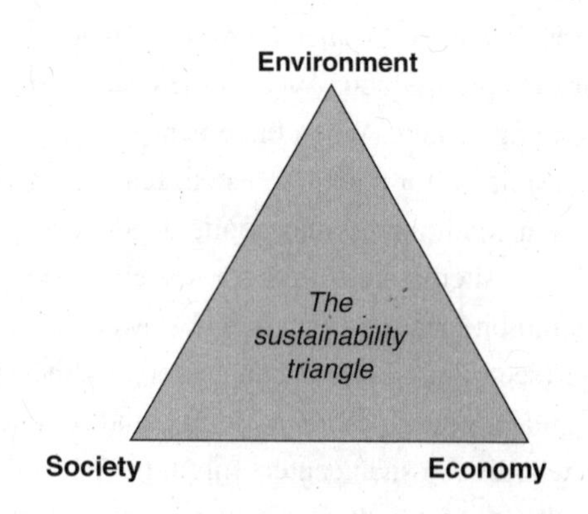


Figure 2-3 *The concept of sustainable development.*

# Project 1: Estimating Your Existing Transportation Related Emissions

## You will need

- A pen and paper.
- A calculator, maybe.

## Steps

1. It is a good idea to find out where we are now, so that we know that we got somewhere, when we get there. Start by taking account of all the travel that you do. Note the distance you travel and the frequency of your trips, as well as what mode of transport you use (how you travel). The following list might help:
   - Travel to office or school ____ miles per week, by ____.
   - Travel for groceries ____ miles per week, by ____.
   - Travel for entertainment ____ miles per week, by ____.
   - Travel for vacations ____ miles per week, by ____.
2. Next, use these very rough estimates (or research your own) to assign emission factors to your current travel habits:
   - Travel by mid–large-sized gasoline-powered passenger car—500 g (1.1 lb) of $CO_2$ equivalent per mile traveled (or around 300 g or 0.7 lb per km). These are rounded up to try to get an easy number. Variation can be large: e.g., an efficient car could be as low as 200 g per mile (~125 g per km). If you have a smaller car, you could adjust this to better reflect your circumstances. The emissions figure is for the car, so if you travel with one or more passengers for many trips, you could reduce the number proportionally for those trips.
   - Travel by plane—start with an estimate of 500 g (about 1.1 lb) of $CO_2$ equivalent per person per mile traveled (around 300 g or 0.7 lb per km). Emissions from airplanes are contentious, because many people believe that emissions made high in the atmosphere have a greater effect than emissions at ground level. This effect is called radiative forcing and is thought to multiply the effect of emissions by a factor of 2 or more. Emissions from short flights (like short car rides) are also considered more polluting and less efficient. Of course the actual emissions from your flight will depend on things beyond your control (like how many people are on the flight), but this ballpark figure will get us started.

### Tip

Use the following conversion figures when looking up emissions data for your particular vehicle:
1.6 km (kilometers) = 1 mile;
1 lb (pound) = 454 g (grams).

   - Bus and train travel—100 g (0.2 lb) per mile (62 g or 0.1 lb per km) of bus or train travel is again rounded up. This also depends on several factors beyond your control, such as how full the bus is and what fuel the locality you live in has decided to use. For example, the city of Calgary, Canada powers its urban train system entirely through wind power, so it would not have direct emissions associated with it. If you take a boat or ferry on a regular basis, you could use this factor as a rough estimate.

- Walking and cycling have very few emissions associated with them, so it is safe to leave them out from the carbon impact calculations. They create a large number of social, health, and environmental benefits though, so don't forget them.

3. Next, add up all of the travel you do for various reasons, by each mode of transport. As an example, let's look at a person who travels 20 miles roundtrip to work 5 days a week, but does their grocery shopping locally and only goes to the movie theater 5 miles away for entertainment once a week, sharing a ride with a friend. Assume also that the person takes two yearly vacations on a long-haul flight—assume these are 2000 miles (3200 km) each way—and six short flights for work—estimated at 1000 miles (1600 km) each way—per year. The weekly travel total looks something like this:
   - Car trips: 100 miles for work (5 days × 20 miles per day) + 5 miles for entertainment (10 miles/2 passengers) = 105 miles of weekly driving-related emissions
   - Plane trips: 8000 miles in long-distance trips (four flights × 2000 miles each) per year and 12,000 miles in short flights (12 × 1000 miles each) = 20,000 miles of airplane emissions per year. Divide by 52 weeks in the year to get a figure of 384 miles of weekly (equivalent) airplane-related travel emissions.
   - No walking or cycling emissions from collecting groceries.
   - No bus or train emissions.
4. Finally, multiply your weekly travel emissions totals by the emissions conversion figures:
   - Car trips: 105 miles per week × 500 g/mile = 52,500 g of $CO_2$ equivalent per week. Divide by 1000 for kilograms and multiply by 52 weeks for a yearly estimate of 2730 kg of $CO_2$ from car travel.
   - Plane trips: 384 miles per week equivalent or 20,000 miles per year multiplied by the 500 g/mile figure (ignore radiative forcing for the moment) results in 10,000 kg of $CO_2$ from air travel per year.
   - Find the yearly total of your car and plane emissions: 10,000 kg + 2730 kg = 12,730 kg $CO_2$.
   - No emissions from other modes. In our example, the yearly total of 12,730 kg is nearly four-fifths air travel (from eight flights) and one-fifth car travel (over 90% is travel to and from work).
5. Think twice before driving to the store next time, and three times before heading for the airport. Being green does not necessarily mean not flying, but it does mean being conscious of these things, and adjusting your behavior where possible.

## Locked-in

It is common for people who own a car to use it for the vast majority of their activities. Part of the reason for this is economic, because once the cost of purchasing and insuring a car for your exclusive use has been paid for, the increased cost to you of using a car for the minute-long drive to the store is very small. Among the problems from a pollution perspective is that short trips by car tend to have the highest emissions of all trips, and make up almost three-quarters of all car trips in North America. Cars run more efficiently and have fewer emissions when they are warm, but using a car for short trips (less than about 5 km or 3 miles) doesn't give it the chance to warm up. The economic reality of owning a car tends to work against its most efficient use, but if gas prices rise, we may see that change. Car clubs are another way of changing the financial incentive to take short trips, because they tend to charge a per kilometer (or per mile) charge that reflects all of the costs that go into running a car while the member is using it. Car clubs can have their own host of problems and are not suitable for everyone, but they are one alternative, among many, to the personal automobile.

# Project 2: Finding Another Way Around

Reducing our collective emissions is going to involve a lot of creative thinking. This project challenges readers to try a new way of getting around their city, without a car. Maybe this means taking the bus into work one day, or taking your bike to the store, or maybe taking the extra time to walk with your kids to school. What is important for this project is for readers to understand that there are very often several alternatives to taking a car. There are certainly times when it is necessary, and these will be different to everyone, but it is far too easy (and common) for complacency to become the dominant motive behind short car trips.

## You will need

- A day to try something new. Try to make it a day that is not stressful in its own right. Communicate to those around you (your boss, family, etc.) what you are trying and why. Ask them to support you by being understanding: things might not go like clockwork the first time.
- Maps of the bus routes in your area, or possibly maps of walking and cycling routes from your home to office, school or the shops. You may be surprised to find that there are cycling and walking routes that, though not necessarily direct, would allow non-automobile traffic to travel in relative comfort and undisturbed by the air and noise pollution of automobiles. Often these paths are not obvious to accustomed drivers, so be prepared to do a little bit of looking around. Talk to others in your area, as many trails and networks in cities and the suburbs are informal routes rather than designated cycle paths. These follow quieter streets and alleys rather than riding alongside traffic and can be safer for the inexperienced biker.
- If there are no non-automobile alternatives, try carpooling. See if you can find others in your neighborhood and community making similar journeys to yourself and try to combine trips. You could save money and make a new friend.
- A pen and paper (to record and later communicate your experiences); a computer could work equally well.
- Appropriate tools for your new travel journey: a drink and change of clothes if you are riding to work; maybe a magazine if you are taking the bus or train for the first time (you'll be surprised how much time you have if you're not driving).
- If you don't drive, maybe you fly. Think hard about whether you could take a bus or train, or avoid the trip altogether. Because airplanes are able to cover great distances in a short time, flying has some of the highest per trip emissions of all forms of travel, even though people only tend to take a few flights per year.

## Steps

1. Take it slow. Assess your options and their relative benefits before deciding one way or another. There might be a bus route you don't know of, a cycle path that goes to the door of your office, or a neighbor driving by your office everyday.
2. If you don't succeed at first, try again, then try something else. If there are no buses near you, perhaps you could drive part of the way, or start talking to the local government and transit providers to see why there is not transit and whether that could change. If the weather is a problem for the prospective cyclist, look at later projects in this book, where we build weather protection for human-powered vehicles.
3. Try your new mode one day per week to start. When that becomes a habit, move onto bigger things. Remember to move onto bigger things; it can be easy to become complacent after starting to ride your bike once a week in the summer. That is a start, but not a solution.

4. Once you have a routine that better reflects your concerns, try making a renewed effort at estimating your personal carbon emissions. You could be surprised at how much you have reduced them without introducing hardly any "new" technology at all.

## DIY or not, just get it done

When I was 11 years old, I got my first new bike for my birthday. It was a white BMX bike and it worked perfectly from the day I got it, until it was stolen not long after. I didn't get to go back for another new bike, but instead went to where our family had always purchased our bikes: the City of Edmonton police auction. The police auction sold all varieties of bikes that had been found abandoned in the city or confiscated for whatever reason. The bikes were never in very good shape, and always needed some repairs before I could ride them. As a young evil genius, I thought this was half the fun. My dad would take us down to the Edmonton Bicycle Commuters Club, where for a small yearly membership fee we could use their wonderful bike garage and get advice from expert bike mechanics on how to fix your own bike. Yep, this was (and still is) a place where people can come to fix their own bikes, rather than have someone else do it for them. The skills I learnt there have been some of the most valuable of my life, and I have consistently been astonished that not everyone in the world has had the experience of straightening a bent frame, nevermind fixing a flat tire on their bicycle.

I get the same sort of astonishment when I see the number of bicycles in various states of disrepair sitting in people's back yards. It occurs to me when I start to talk to people that even basic bicycle mechanic skills are not commonly available. If we are going to start using bicycles more often, something needs to be done about that. Which doesn't mean that everyone should be able to swap a rear derailleur in 40 seconds or less, because the reality of going on longer-distance bike rides with experienced cyclists is that not all of them can fix a flat tire either. Some people in larger centers have realized that cyclists are not mechanics, and are reacting. For example, a group of Torontonians is (or at least were) offering a sort of emergency bike-side assistance for city riders. For a small fee, they are dispatched by a phone call, with tools to repair your bike, and a trailer to tow it away if they can't fix it right away. This eliminates the inconvenience of bicycle maintenance for those who are less technically proficient, and is a tremendous step on the road to getting more people riding their bicycles. For the real evil genius, I don't expect technical proficiency to be an issue, but a few hints and notes on what to do couldn't hurt. This is, after all, the first step in getting the most efficient vehicle on the road moving. And possibly, the start of a series of emergency bicycle repair services across the country (wink-wink, nudge-nudge).

# Project 3: Flat Isn't Phat

## You will need

- A flat tire on your favorite bicycle (or one of your best friend's, your uncle's or little sister's).
- A deluxe patch kit from your local bike store that contains at minimum the following parts (see Figure 2-4 for details of what you are looking for):
  - three plastic tire irons (yes, you should actually use iron tire irons, but they are getting harder to find)
  - rubber patches of various sizes
  - rubber cement
  - small metal scraping tool.

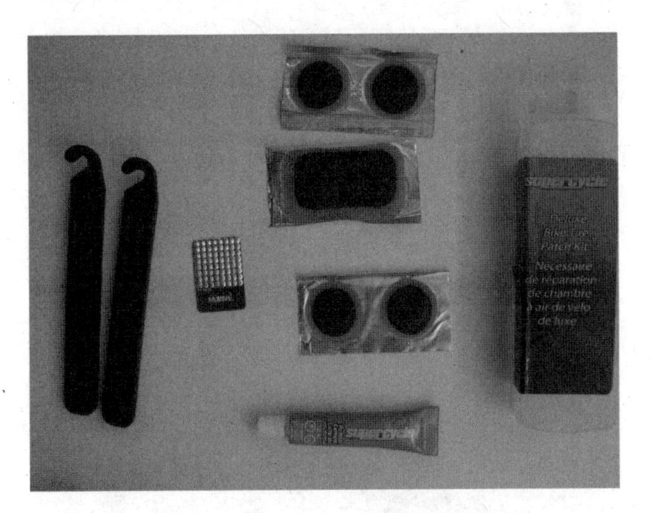

Figure 2-4 *A typical deluxe bicycle tube repair kit.*

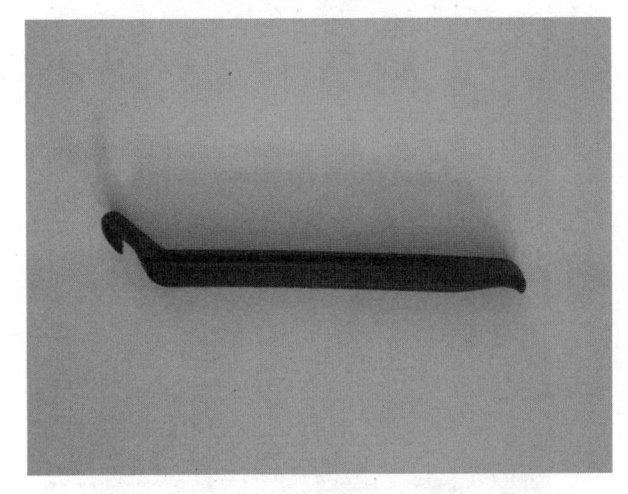

Figure 2-5 *Detail of a common tire iron (now made out of plastic), showing the end (pointing down) that interacts with the tire.*

- A bucket (or sink) filled with water.
- A bicycle air pump.

## Steps

1. The first thing to do is to flip your bike upside down (or raise the tire if it is a three or more-wheel vehicle). For a bicycle, this avoids scratching the rims on the ground once you take the tire off, and makes the bike relatively stable while you work on the flat tire.
2. Next, gently push the tire back from the rim at one point.
3. Using the end of the tire iron that has the bend in it already (see Figure 2-5; not the end with the hook), push the tire first away from the rim (toward the far edge) then lift up (see Figure 2-6).
4. You should be able to pull the tire far enough over the rim to enable you to hook the other end of the tire iron to one of the spokes (see Figure 2-7). Hook it there and leave it.
5. Next, take a second tire iron and do the same thing about 15 cm (6 inches) along the rim (see Figure 2-8). If it is tight, try closer to the first tire iron, but as far apart as you can manage. The two tire irons should both be pulling the tire over the edge of the rim.
6. Use the third tire iron in the same manner on one side of the other two irons, about the same distance away as before. Using three tire irons is best: if you only have two and it is not a tight tire, it is often possible to remove one of the two tire irons and insert it a little further along without the tire sliding back onto the rim. You may also be able to simply slide the second tire iron along to the rim to free the tire.
7. When you insert the third tire iron, the middle one should start to become free of the rim.

Figure 2-6 *Pushing the flat end of the tire iron under the edge of the tire.*

**Figure 2-7** *Hook the tire iron to the spokes once the flat side is inserted under the tire.*

8. Pick it up and insert it under the edge of the tire further along the rim from either of the other two. Now the tire should be loose enough that you can slide the tire iron around the entire rim and pull one edge of the tire off (see Figure 2-9).

**Figure 2-8** *Insert a second tire iron; then, slide it along the rim to free one edge of the tire.*

**Figure 2-9** *Pull the tire back to reveal the inner tube.*

9. Reach into the tire and feel the inner tube, find the valve (see Figure 2-10) and gently pull it out of the hole in the rim. Then you can remove the tube from inside the tire (Figure 2-11). There is no need to take the tire entirely off the rim, just leave it hanging half on (unless in the case pictured, your bicycle is actually a recumbent tricycle, shown in more detail in a later chapter).

**Figure 2-10** *Detail of the valve; push it back into the rim before trying to remove the inner tube.*

Figure 2-11 *Pull the inner tube from the tire.*

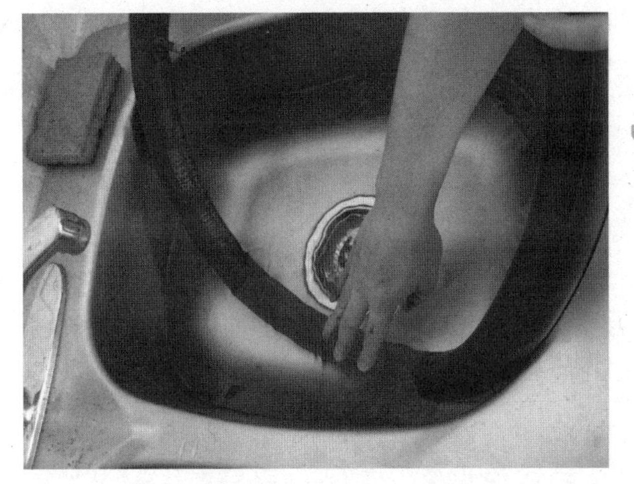

Figure 2-13 *Place the inner tube in a bucket or a sink with some clear water in it. Look for bubbles, which indicate where there is a leak. Mark the spot once you find it.*

10. Your inner tube won't be completely free of the bike unless you took the wheel off. The front or rear fork will keep it attached, but you should be able to fix it without removing the wheel from the frame.
11. With your air pump, fill the inner tube with a small amount of air, until the tube gets its circular shape back (see Figure 2-12). You don't want to put too much air in or it will get hard to maneuver the tube while it is still attached to the bike.

Figure 2-12 *Filling air into the inner tube with a hand pump.*

12. Run your hand around the inner tube to see if you can feel any air leaks. If you do, make a note of the location by keeping your thumb on it. Keep checking to see if there are more leaks.
13. If you don't find any air leaks with your hand, grab the bucket of water and put the tube (bit by bit) into the bucket (or a sink, see Figure 2-13). If you see any bubbles coming out of the tube, you have found your leak!
14. Keep track of the hole with your thumb and dry the inner tube a bit. Scrape the area around your puncture with the metal tool provided (see Figure 2-14). This cleans the area and gives the rubber cement something to grip onto. Don't be too rough because you could create another puncture.
15. Find a patch that is big enough to completely cover the puncture and peel the plastic covering off of it (Figure 2-15).
16. Apply some rubber cement to the tube and to the underside of the patch (Figure 2-16).
17. Wait a minute for the rubber cement to cure a little before your press the inner tube and patch together.

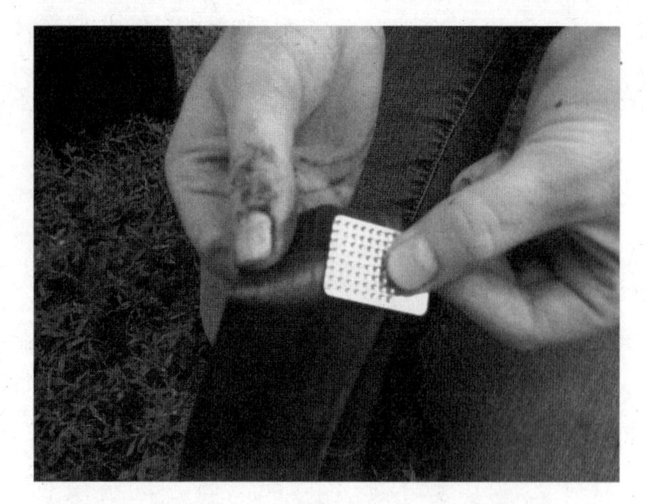

Figure 2-14 *Scrape the tube gently, to clean the surface and give the glue a rough area to adhere to. Be careful, as it is possible to create a bigger puncture at this stage.*

Figure 2-16 *Apply some glue to the tube over the patch you previously scraped.*

18. Place the patch onto the inner tube and press down (Figure 2-17). If some of the rubber cement squeezes out the sides, spread it around with your thumb. Spread a little extra rubber cement around the edges of the patch to make sure that there are no air leaks out the sides.
19. Let it dry for an hour or more (Figure 2-18). You can check the condition of your tire and rim while you are waiting. Look at the inside of the rim. Your rim should have a thin strip of rubber upon which the tube sits. Oftentimes, punctures are caused by one of the spokes pressing through this rubber layer and poking through the tube. Repair or replace the rubber strip if it is damaged, and look for other foreign objects in the rim and tire.
20. Once the inner tube is dry, you'll want to briefly fill it up with air and run it through the bucket to make sure you got all the leaks.
21. Then, to begin putting your inner tube back onto the rim, line up the valve with the hole in the rim and push it through (Figure 2-19).

Figure 2-15 *Peel both sides of protective packaging off of the patch. Start by rolling the patch away from the aluminum backing at one corner and then peel diagonally to avoid tearing the patch.*

Figure 2-17 *Press the patch onto the inner tube and hold for a second. If some glue leaks out the sides, spread it around the edges and make sure they are securely stuck to the tube. The edge of the patch is a common place for leaks to reoccur.*

Figure 2-18 *The completed patch job drying in the sun.*

22. Place the inner tube inside the tire (Figure 2-20) and then start pushing the tire back onto the rim.
23. When you get about three-quarters of the way around the rim, it might start to feel difficult to stretch the tire over the rim (Figure 2-21). You may be tempted to reach for the tire irons that were so handy when removing the tire. Don't do it! It is very easy to puncture your tube again when putting the tire on.
24. Despite what you might think at this point, it *is* possible to put the tire back on without using any tools. This is the safest and correct way to put a tire back on the rim. Grab the back-side of the rim with the fingers of both hands and use your thumbs in a rolling action to push the tire back onto the rim (Figure 2-22), little by little.

Figure 2-19 *Feed the valve stem on the inner tube through the hole on the rim.*

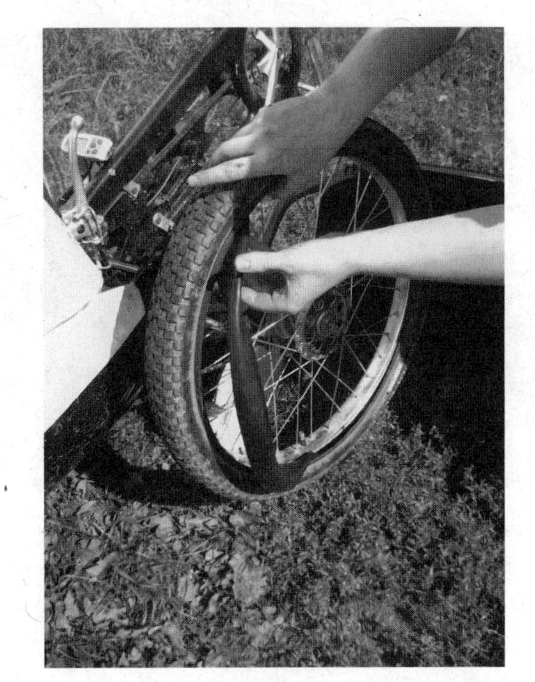

Figure 2-20 *Place the inner tube back into the tire.*

Just start at one end and keep working down until the entire tire has squeezed itself back on the rim.

25. Grab your air pump, fill up your tire (Figure 2-23), and test out your repair job.

Figure 2-21 *With the newly repaired inner tube inside, start pushing the tire back onto the rim.*

**Figure 2-22** *Use your thumbs in a rolling action to coax the last parts of the tire onto the rim. No tools are needed at this point!*

**Figure 2-23** *Fill the tire with air, and check the pressure (read on to the next project if you are unsure how to do this).*

## Driving smart

Reducing the amount that each of us travels, particularly by car and airplane, can go a long way towards using energy efficiently. There are, however, many situations that today are too difficult to consider not using a vehicle for, particularly in the North American context. For this reason it is important to consider the efficiency of the vehicle(s) you drive. There are many things that affect how efficiently you use energy while driving. Occupancy, or how many people there are in the car, is obviously one of the most obvious ways of increasing the efficiency of fuel use around the world. There are simply far too many single-occupant vehicles traveling around our roads and highways today. Carrying four empty seats is simply not smart; try to fill them up when you can. Some cities encourage this by providing carpool lanes for higher-occupant vehicles. If these aren't in place in your city, ask why not? If your government representative can't answer your questions, ask a reporter. They may find a story in it, which will help bring the environment even more into the public domain and carpool lanes to your highway.

# Project 4: Round and Round

Just before you have a full car rolling down the road, it is important to consider the state of your vehicle. Questions you should ask about the state of your car include: Has the car been maintained lately? Was the air filter and oil changed? Did I check the tire pressure? Do I even know where to find the correct information? Each of these details about your car can have a significant effect on the fuel economy and performance of your vehicle. For example, studies show that about one in three vehicles on the road have at least one tire under-inflated. One under-inflated tire has been shown to reduce fuel efficiency by up to 5%, but only takes a few minutes to correct. Poorly maintained tires will also cost more money in tire maintenance, and could be dangerous. Also, 5% of your fuel bill goes up in smoke because of too little free air. That doesn't seem very ingenious to me, so checking your tire pressure once a week is highly recommended. If you don't know how to check your tire pressure—nevermind where to find the correct pressure to fill it to—you are not alone.

## You will need

- Your reading glasses (maybe even a magnifying glass).
- A tire gauge and potentially an air pump (the one at your gas station will do).

## Steps

1. The first step is to find the correct tire pressure for your vehicle.
2. If you went to look at the tire itself, you are wrong, sorry. Tires typically have the maximum safe pressure rating stamped on their side. This is not necessarily the appropriate pressure for your specific vehicle.
3. The correct tire pressure is typically found on a small metal sticker affixed to the frame on the edge of the driver's side door (see Figure 2-24). Alternatively, look in your owner's manual. If that sounds like a lot of work, think about what five percent of your fuel bill could buy.
4. Now that we know the correct pressure we're looking for, we can check each tire individually. Remove the dust cap from the valve stem and press the tire gauge flat onto the stem (see Figure 2-25).

**Figure 2-24** *Location of the label on the inside of the car door, showing the correct tire pressure for the vehicle.*

5. If the pressure indicates that the tire is low, fill it by pressing the nozzle straight down onto the stem, just like we did with the pressure gauge. Some compressors require a handle to be pushed to start the air.

**Figure 2-25** *Using a tire gauge to check the correct pressure.*

# Project 5: Driving Smart

Real evil geniuses will know everything they can possibly do to increase the efficiency of their energy use, but sometimes even the most efficient of us can't help but drive. There are, of course, heaps of other things that the average driver can do to squeeze ever more miles out of each gallon of gas.

## You will need

- You (the driver) and a car.
- Some fuel in your car.

## Steps

1. Turn your car off when you are not moving. Idling your car for more than 10 seconds wastes gas and gets you nowhere. The average car in North America idles for about 10 minutes a week, not counting time spent idling in traffic. The least a green genius could do is to turn off the engine while waiting to pick up friends. In many European countries, drivers are in the habit of turning off their engine while waiting at particularly long traffic lights. The lights there even indicate (through an advance amber) that the light is about to turn green, to give drivers a chance to start their engine. It reminds me of a Formula One race, and I like it.
2. Carpool. We mentioned this already, but it is so important that there is no issue doing it again. When you travel alone in your car, you are using it at less than a quarter of its possible (per passenger) efficiency.
3. Buy the smallest and most efficient vehicle that is appropriate to your needs. Car companies post the mileage of their vehicles, so it is up to us as consumers to pay attention and buy fuel-efficient vehicles. Not everyone will find a hybrid suitable (maybe only those who spend a lot of time in urban areas) and not everyone can live without a truck either. What is important is choosing the most efficient vehicle among those that fit your needs, and then using it only when appropriate. Not really genius, just smart.
4. Slow down. Speeding wastes an enormous amount of gas, and doesn't tend to get us places much faster. Modern vehicles are said to travel most efficiently at around 56 miles (90 km) per hour. Test it out: this one is easy.
5. Learn about and practice efficient driving styles. Individual driving habits can result in up to 40 percent more fuel being used for the same trip. Efficient driving techniques include accelerating slowly off of stoplights, never speeding, not accelerating up hills, and reducing the amount of braking needed by looking ahead in traffic. Search the Internet for "hyper-milling" to get started.

# Chapter 3

# Staying Inside

When we are not in our cars, many of us spend the vast majority of our lives inside buildings: often without giving them too much thought. They are protection from rain, snow or sunshine, and we expect them to be warm in the winter and cold in the summer. Often monuments of our achievements, they are frequently things of extreme beauty. Different buildings achieve their varied purposes in different ways and, because we have been building for centuries, there are many types to choose from.

Energy is used in all parts of the building cycle: during construction, in use, and at the end of its life. In addition, the occupants of a building will usually consume energy traveling to and from the location, and their activities while in the building typically have their own energy requirements that must be provided to the site. An environmentally sensitive building should minimize the amount of energy that goes into a building during each phase. We mentioned in the previous chapter that transportation accounts for about half of the energy consumption of the typical North American family. Space heating and cooling (using our buildings) then accounts for a little more than half of what is left, or a quarter of the total energy consumed. Like other places where energy is used, we can account for it. Because they are the next largest consumer of energy, the evil genius who is serious about reducing energy use will need to have an understanding that buildings can be heated and cooled.

Building design has a large effect on the amount of energy consumed in operation during its lifetime. The process of accounting for the energy used in constructing a building has been well developed in the building industry. In the United States this has resulted in the creation of the Leadership in Energy and Environmental Design (LEED) program that offers different levels of certification depending on how many of the program's criteria the structure meets. They are great places to get a feel for the broad range of issues you'll have to consider if you want to reduce the environmental impact of a new building or a space you are occupying. There are also a number of ways that energy-efficient technologies and habits can lower the ongoing energy costs of a building that is already in use. These include altering occupant behaviors as well as retrofitting older buildings to improve their energy efficiency and adjusting to the changing occupant needs and habits.

## Around the house

Our homes are often the buildings that we have the most control over, and are therefore the places where we can begin to make changes in our energy consumption patterns with the most ease. It is useful to think about how we can reduce energy consumption there first, and then move on to other areas, such as work and community.

The air around us plays the biggest part in keeping our homes habitable and comfortable to live in. Humans need oxygenated fresh air circulated around us pretty much all the time. We prefer it to be about 20° C (68° F) for the most part, though we can tolerate a larger range of temperatures, with some adjustment. Allowing for a greater amount of fluctuation in the range of temperatures that are considered acceptable can have a significant effect on the amount of energy that a building consumes. This is particularly relevant in the shoulder seasons, spring and fall, when heating and cooling systems are being switched over. On a day where the outside temperature is a seasonal 16° C (60° F), if the inside thermostat is set to 18° C (64° F) exactly, the heating system is using energy that day. If, instead, a 2° C leeway on either side of that temperature is allowed, the day passes without any energy expended on heat.

During the rest of the year, lowering the usual temperature by as little as a few degrees can lower your bill by 10 percent or more. It has long been a mantra of the environmental movement to save energy by putting on another sweater (or jumper) instead of raising the thermostat. Altering everyday behavior certainly does have a part to play in reducing energy consumption, especially if it is the beginning of a more substantial change. Dealing with a slightly colder room with a sweater might be OK today, but increasing the insulation in your home or a more efficient furnace are longer-term goals that such small actions should be progressing toward.

## Long- and short-term actions

Many aspects of life, from small everyday things, through to large important purchases, have an environmental aspect to them. It is useful to get into the habit of thinking about energy and its use over a long period of time when making all range of decisions and adjustments. For example, if you are sitting in a cold room right now, the first step to reducing your energy consumption might be to put on a sweater rather than turn up the thermostat. But in the medium term your goal should be to look into where energy is being lost or used inefficiently in your house. It could be that the insulation in the roof, walls, and the windows all combine to let a lot of your heat out. A short-term improvement could be to install a plastic double-pane window for about five dollars. A medium-term idea could be to look at improving your loft insulation, then replacing leaky windows (see Figure 3-1). Often, these will pay for themselves in reduced fuel bills over a short period of time.

Over a slightly longer period it might be worthwhile to look at replacing the furnace with a more efficient one. Over the even longer term, it is probably wise to look into collecting solar thermal energy through retrofitting your existing property, or building a passive solar house. In cold parts of Europe, zero energy homes are possible and becoming common, using only the sun's heat to maintain a comfortable temperature all year round. These houses take full advantage of the sun's energy and are super-insulated to maintain their temperature. It is not a dream: it is now. But for most of us, it is the future.

**Figure 3-1** *Old single-pane windows with many gaps around the edges visible. Keep reading for both long- and short-term improvements if they look familiar to you.*

In the next few projects, we build crude models of a house (OK it's just a room; but feel free to expand), in order to demonstrate how hot and cold move around. We do it this way rather than point you at your house right away, because there is a lot of value to understanding the concepts behind why something works rather than just doing it.

# Project 6: Fabric Walls



Little things can make a huge difference. My mother used to tell me to close my curtains at night, but I never understood why. When I moved out on my own, it was not until I started to pay the heating bills living in a drafty room that I figured it out. It was to keep the heat in on those frigid –30° C (–22° F) nights. Fabric walls, or curtains, are a useful and inexpensive energy-saving tool, but can also be effective in short- and long-term energy conservation strategies. Let's prove it, so you'll never forget to close them again (because let's face it, I still forget all the time). If you're like me, you'll be inspired to dream of automatic systems that rely less on human intervention to save energy. These exist in commercial form already, but are primarily available for large construction projects rather than a home scale or are focused on security rather than making energy savings.

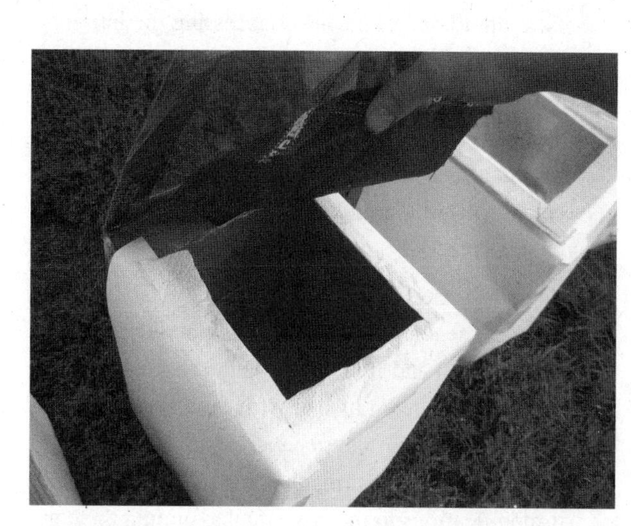

Figure 3-2 *A medium-sized test box under construction.*

## You will need

- Four boxes or cubes, to imitate the four walls of your house. They should all be the same in size and material. You could use wood to make cubes, or find cardboard boxes, or something else that is handy. Large "test cells" would ideally be around 1 meter (or 40 inches) square, and a scaled-down version could be closer to 10–15 cm (4–6 inches). A medium-sized "test box" is shown in Figure 3-2.
- We will cut a square hole about half a meter across on one side of three of the cubes, so make sure you can cut through the material. See Figure 3-3 for a diagram of what we are trying to achieve.
- Four panes of glass, each a little bigger than the hole you'll cut in the boxes. An alternative is to use Plexiglas, or even plastic wrap (film) from the kitchen, particularly if making a smaller version (though this gets tricky with the last variation).
- Several window coverings, each at least as big as the glass. In an ideal world you'll have:
    - a set of light curtains
    - a dark heavy fabric.
- Four hot water bottles or reused plastic water bottles (as in the example) but be careful to use only warm water as boiling water will collapse the plastic.
- A thermometer (digital oven thermometers work well) or four to avoid moving the thermometer between boxes.

Figure 3-3 *Diagram showing where to cut the box.*

## Steps

1. Cut a square hole in one face of three of the four cubes, removing about half of the wall. You will also want to have a means of accessing the interior of your crudely modeled home, probably through the top. The holes you just cut are the windows of our "model home." Alternatively, create an external frame window and fix it to one side, replacing a wall so it can double as an access point.
2. Mount windows into three of the boxes. Try to get as good a seal around the edges as possible; thick tape is useful and if strong enough can hold the window in place on its own. It is very important that the fit is about the same among all four boxes, as you will be comparing the thermal performance of modifications to the boxes and assuming they are otherwise the same.
3. Through the opening, mount curtains inside two of the windows (see Figure 3-4). Leave the third one without a curtain
4. Fill up your hot water bottles with boiling water, all at the same temperature (you can take the temperature to be sure, see Figure 3-5).
5. Place one hot water bottle in each of the cubes; make sure that in the ones with curtains they stay closed (see Figure 3-6).

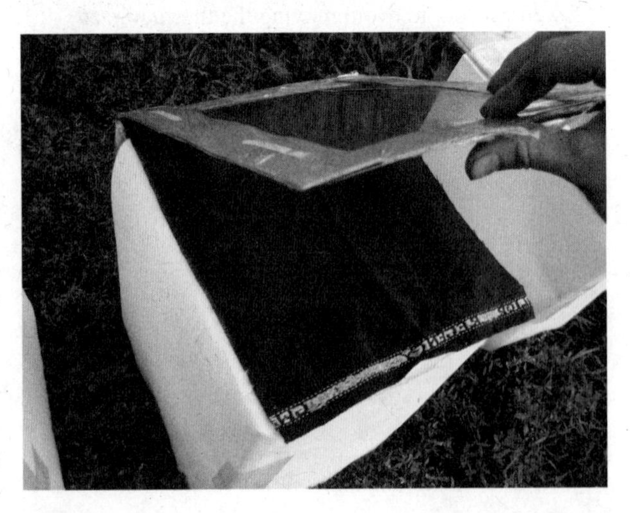

**Figure 3-4** *A thick curtain being mounted behind the plastic wrap "window."*

**Figure 3-5** *Taking the temperature of the warm water with a digital oven thermometer before it is put into the "model room" with a light curtain.*

6. Leave outside for a few hours *at night or in the evening as the sun wanes* (not bright sun or the heat of the day—very important as we are exploring heat loss).
7. After several hours (it will vary depending on what you used to make your box out of, where you are in the world, and what season it is) record the temperatures of the water inside each of the four hot water bottles.

**Figure 3-6** *Four model rooms with reused hot water bottle "heat sources" inside, cooling down in the waning sun. The model room on the far left has no curtains, to its right is the model room with no window, followed by a thick curtain in the window, and a thin curtain (kitchen towel) in the last window. All windows are made from plastic wrap, on a cardboard frame mounted on identical cardboard boxes.*

8. The cube with no window at all should be the warmest. The cube with the thick curtains should be next, followed by the thin curtain and finally no curtain at all. If you don't see a great enough difference between the four, try shortening the time you leave the water bottles out.

## Variations for hot climates and seasons

1. A variation, for hot climates or seasons, is to place the boxes in the sun with the curtains closed but with a cold water bottle, and check which bottle takes the longest to heat up. In many hot areas it is equally important to remember to keep the drapes closed during the day to reduce the amount of solar heat that comes into spaces that we would like cool (see Figure 3-7).
2. Experiment with different window coverings, both internal and external, and also try shortening the curtains and lengthening them. External shutters are common in hot climates, ever wonder why? (see Figures 3-8 and 3-9).

**Figure 3-7** *Curtains being used to shade the sun, and let in light, in the hot season.*

**Figure 3-8** *External shutters in a hot country.*

3. Try covering one of the windows with a second layer of clear material, like plastic wrap. You can easily build a square frame slightly bigger than your window, stretch the plastic wrap over it to make a clear frame, and then affix it (preferably) on the inside of one box. Repeat the tests to see what the impact of adding an inexpensive double pane can be.

**Figure 3-9** *Roll down shutters can be automated to close at certain times.*

## Habits and upgrades

The last experiment hopefully encouraged the reader to take his or her bike out for a spin to pick up some new heavyweight curtains for the winter time, at least. You should have proved the impact that simple, inexpensive tools, and little changes of behavior can have on the energy needed to moderate the temperature in a poorly insulated space. The next step is to put those ideas into practice around the house. Start by closing your drapes at night (or the day if you're in a hot climate) and make sure they hang all the way down to the ground (you can test the effect of this using your test boxes too). But even further down the line, let's not forget that closing and opening the drapes at a particular time, and adjusting other ways in which a building can operate to save money, don't have to be done by people everyday if they are thought about at the outset. New eco-friendly buildings often incorporate automatic vents or windows that can open on hot days, and an insulative blanket to pull down in the night. These technologies mostly haven't made it into the housing market yet, but watch for them and ask about options if you are looking to buy or build.

The next relatively painless action you could try is to lower your thermostat slightly for a few days, and see if it is still acceptable to you. If you find it quite cold and have a drafty home, draft proofing strips are available at most hardware stores and are very easy to install (see Figure 3-10). Leaky door and windows are some of the worst culprits for energy waste. If you are serious about reducing the energy that your house consumes, you might at some point hire an energy assessor. In many countries these are technicians licensed by the government and trained, among other things, to do a blower test. In this test, a large fan is attached to the door and air is pumped into the house. An airtight, efficient house should have a pressure build-up during the blower test and a less airtight house would have less pressure build-up

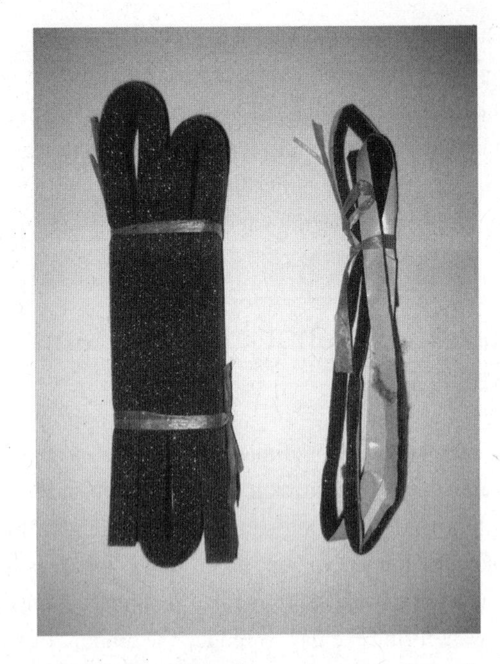

**Figure 3-10** *Draft proofing strips of various sizes, with adhesive tape and without.*

## Windows

Leaks are often found around windows in old buildings. Before our ability improved to make inexpensive double-pane glass, the vast majority of buildings were constructed with just single-pane glass. The problem with this, as we saw in the last project, is that glass is not a terribly good insulator. A double-pane of glass overcomes this limitation by sandwiching a layer of air or other gas in between two layers of glass (see Figure 3-11). Air is a fantastic insulator, among the best we know of.
The problem with using air is that it tends to seep away from where you want it to stay, and mix with the other cooler or hotter air, and so insulation tries to keep that air sealed. Older-style double-pane windows used two individual panes of glass, on two separate tracks, which kept a layer of air sealed in between them while the windows were closed. They are better than single-pane glass, but worse than modern sealed double-pane windows, which trap air (and more recently other gases) in between two layers at the factory (see Figure 3-12).

Figure 3-11 *Newer-style airtight double-pane windows.*

Figure 3-12 *Sealed double-pane glass windows awaiting assembly in a factory. A layer of gas is trapped in between panes of glass at the factory, resulting in significant energy savings compared with a single-pane window.*

## Project 7: Fixing Your Windows

Not all of us are in a position to be thinking about replacing the windows of our house. It is, after all, an expensive proposition that usually requires a professional installer to get a good seal and fit. A particularly attractive alternative, if on a budget, is to make your own double-pane windows. We did a small version in the last project and saw the impact that just a small layer of Clingfilm and a layer of air had on the heat retained in our boxes (or kept out). The layer of plastic tends to block air leaks, and the layer of air trapped in between the glass and plastic becomes an insulating layer.

### You will need

- To buy the specially designed packages that come with heat-sensitive plastic and double-sided tape (your local hardware store is bound to have them).
- An alternative is to buy a roll of double-sided tape, and a roll of heat-shrink plastic (ask to make sure).
- Ordinary plastic stretched over a large frame works on a large scale as well, and is useful if you have

a large window or a sliding door that leaks. You will want some draft proofing pieces and probably nails to hold your makeshift window in place on the inside or outside (either can work).

## Steps

- Prepare for installation by cleaning the windowsill of dirt and dust.
- Stick one side of the double-sided tape to your windowsill and peel off the other layer protecting the second sticky side.
- Place the plastic covering along the tape, starting at the top corner and pulling it as tight as you can while pressing it along the rest of the plastic tape.
- Use a hairdryer to shrink the plastic over the glass and tight to the frame.

## Variations

- To test the value of adding a layer of plastic on a small scale is fairly easy with our test boxes. Simply build a simple frame with a stretch plastic cover over it, then affix it to the window with a gap of air between the two. We also explore this effect on a small scale with a solar oven later on.

# Project 8: Using Thermal Mass

Once all the drafts are sealed, and the windows double-paned, it is time to start looking deeper into the building for places to save energy. We saw already that a layer of air in between two panes of glass helps maintain a temperature difference, but there are also other ways of modulating temperature. Some of the buildings and materials around us are bricks and other heavyweight materials (like water), and these behave (from a thermal perspective) quite differently to materials that insulate us. Used together, they can save us an enormous amount of energy (see Figure 3-13). One way is by storing energy (which sometimes arrives when we don't need it, like on a hot sunny day) for when we do (like late at night).

**Figure 3-13** *Thermally massive concrete flooring being installed on top of a layer of insulation in a geothermal (earth as a source of heat) building, described more later in the book.*

## You will need

- Two of the boxes from the earlier project, both with windows.
- Some bricks or stones. You'll want to be able to line the outside wall somewhat, or surround the hot water bottle, so appropriate to the size of your box.
- You thermometer from the last experiment, and the hot water bottles.

## Steps

1. Stack the bricks inside the walls of one of the boxes, and place the hot water bottles in the center of each (see Figure 3-14).
2. Place your test cells outside and take the temperature of the hot water bottles.
3. Leave the boxes for a few hours (see Figure 3-15); then take the temperature of the water again and feel the rocks with your hand. When they feel warm to the touch, go on to the next step.
4. Remove the hot water bottles from both boxes, leave your thermal mass inside, and continue monitoring the temperature.
5. The box without any thermal mass will cool down almost immediately, whereas the box with thermal mass should retain the heat inside our model room for some time.

Figure 3-14 *Thermal mass (large rocks) surrounding the warm water bottle in a "model room."*

Figure 3-15 *"Model rooms" with and without thermal mass inside.*



# Project 9: Testing Methods of Insulating

We have just seen that something with a thermal mass can store and release heat slowly over time. This can be very useful in a variety of situations. Many modern passive solar houses take full advantage of this concept and store the sun's heat during the day to be released at night. In those houses, there is only so much heat from the sun, so they are also very efficient in keeping the heat where it is useful, inside the walls. We also know that thin layers of air can help maintain a temperature difference through glass. In order to keep in scarce heat, the same concept can be applied to the walls and roof.

## You will need

- Two boxes from Project 6.
- Something to use as insulation, such as straw, shredded paper, or fiberglass home insulation. Use the same material to insulate all the boxes, for now (a variation could be to try other types of insulation).

## Steps

1. Place the insulation around the outside of Box 1, and on the inside of Box 2.
2. Place some thermal mass (like some rocks) inside Box 1 surrounding your heat source (water bottle), and surrounding the outside of Box 2.
3. Leave them outside for a few hours. When you come back, take all the hot water bottles out and measure their temperature. Record any difference.
4. The box with insulation inside (Box 2) should have the warmest hot water bottle, but the one with insulation outside (Box 1) should have the warmest air temperature inside (feel it with your hand), and will retain some heat after the heat source (the hot water bottle) is removed.

## Variations

Remove the thermal mass and then remove the insulation from the top of one box. Compare how long it takes for the temperature in the water bottle to drop, to a second case with insulation all around, and a third with insulation left off of one side. Because heat tends to rise, the box without insulation on top should lose heat faster. Think about your attic after this, and go have a look at the insulation up there. Attics are often the most effective places to begin insulating.

## What's going on?

When you place a hot object next to something that has a high thermal mass, like a brick, the brick absorbs some of the heat and stores it. When that heat source is removed, the brick will slowly start to release some of that stored heat into the surrounding environment. When you place insulation outside of the bricks, it slows the heat from radiating out in that direction, so more of it stays inside of the box. In the boxes where you place the insulation on the inside of the bricks, the insulation worked to keep the heat from the bottle close to the bottle, and the bottle should have stayed pretty warm. The bottle would have kept the inside air temperature pretty warm. In contrast, the box with insulation outside the brick would have been colder, because the brick was absorbing heat from the water bottle, which it will continue to release once the bottle is cold or is removed. Insulation outside the brick prevents heat from radiating away.

## Why is this important?

The first step of looking around a house with an eye to efficient heating and cooling is to look at the drafts. Air leaks, as we discussed, are a dreadful waste of heat energy. Next, fixing the windows is usually the most effective and cost-effective way of reducing energy consumption around your house. The householder who has already got this far probably starts looking at the walls and roof as the next step, and they would be right. Insulation can dramatically decrease the amount of energy that must go into heating a given space. Where insulation is placed can have an equally dramatic effect on how the building performs thermally, and should be appropriate to the building's use.

Take the example of a cottage house that is only used infrequently, maybe a weekend or two each month. If the walls of the cottage were brick, and the owner wanted to reduce the amount of fuel that the heating system consumed during each of the weekends they visited, it would not be prudent to place the insulation on the outside of the cottage. The reason is that the brick would tend to act as a heat store for the time that the building was being occupied, because it is only for a couple of days at a time. When the vacationers arrive on a Friday and turn on the heat, the brick walls would immediately start to absorb this heat. Depending on how thick the brick was, it might take a day or more for all the bricks to heat up to a comfortable room temperature, and for most of that time the heating system will have been working very hard to keep the air temperature above the temperature of the walls. After a day or two, the walls would have absorbed all the heat they could, but by then it is Sunday and everyone has left. Before the owner-installed insulation, the walls would have started to slowly radiate heat both outwards and inwards on Sunday when the heat was turned off. This would probably keep a reasonable temperature inside till Monday evening or Tuesday, when no one was around. If the insulation were to be installed on the outside (see Figure 3-16), after 2 days of heating up, the insulation would prevent the stored heat from escaping to the outside. This would result in the inside staying reasonably warm until maybe Wednesday, without any more heat put in. But that hardly does anyone any good, because the vacationers are not there. By the time they arrive the next Friday, the cottage is cold again, and they start the process of heating up the brick walls.

**Figure 3-16** *Insulation and siding being retrofitted over a top of a layer of heavyweight material, on an apartment structure that is constantly heated and occupied.*

The alternative is to insulate the inside of this occasionally used cottage (see Figure 3-17). In that case the insulation would keep the warm air away from the brick and prevent it from warming up in the first place. The cottage would be lightweight and would not take very long to heat up. It would also cool down much faster, but in this case the owner would not be worried about that. If the same cottage were to be used by a family living there full time, then the decision to insulate the outside or inside would be less crucial. A heavyweight building will not fluctuate in temperature as quickly as a lightweight building, so the governing factor in the decision may be the use or habits of the occupants. If this were a space that was occupied throughout the day, for example by a young family with a parent who stays home, then having a thermal mass that takes a long time to heat, like brick walls, or insulation on the outside, would not be an issue, as the house is kept nearly constantly heated. If the same

**Figure 3-17** *Cavities left for the insulation along the inside are visible, along with the thermally massive outside wall. Heat radiating from the thermally massive floor (see Figure 3-13) will be trapped by the insulation, and any heat getting through will first have to heat the brick wall before escaping.*

house belonged to a young couple who both worked, and were only usually home in the evenings, a thermally light building would be more appropriate. The couple could go so far as to install a timed furnace that brought the temperature of the air inside up to a comfortable level right before they arrived home, by turning on a few minutes earlier. A thermally heavy building, on the other hand, might need several hours to raise the temperature and is more likely to have the heat left on throughout the day to prevent it cooling down.

# Chapter 4

# Waste and Value

## Waste to energy

Though we don't often perceive it, much of what we think of as waste is actually a source of energy or, at least, useful for another purpose. Remember always that there are no "waste" products in nature.

For example, when food or feces is left in place to compost, the nutrients from the parts of the food that we don't eat or don't digest gets broken down by bacteria and other living organisms in an aerated compost pile, eventually becoming useful food for a new plant. So long as a pile is aerated (with air) there should be little smell as useful aerobic bacteria will thrive and "fix" (basically store) most of the carbon in the food scraps back into the soil. This is a pretty good use of energy. Too often, the alternative is to have that useful "waste" trucked away (using energy) and thrown into a landfill site far away. Because waste in a landfill tends to have other waste piled on top of it, there isn't a lot of air to help those little organisms in the compost pile break down the waste. Other organisms, which tend to be anaerobic (meaning without air), break it down instead. They do a similar job and also produce a sort of compost when finished; however, they have a side effect, producing methane gas.

Methane gas, as we discussed in Chapter 1, is a powerful greenhouse gas: 20 or so times as potent as $CO_2$. It is also a very valuable fuel source, which some people will be familiar with as natural gas (from a petroleum source), often pumped into our houses and cars and used as a fuel. As a fuel, it is one of our cleanest-burning fossil fuels. As a wasted gas, it is one of the more potent at changing our atmosphere. Sounds like common sense that we ought to use more of it, eh?

Whether we use our food scraps for composting or something else, it seems clear that the last thing we should probably be doing is making methane gas from them, by putting them in landfills and letting gas escape. And some cities and regions are doing their part to capture what is often called "landfill gas," and feed electricity produced from it back into the grid. Other places are doing similar things to another waste product that humans produce a lot of, manure. Manure from animals and humans, when processed anaerobically, produces a similar methane-rich gas.

Several biogas digestors are in operation around the world, from city-sized plants to family-sized pits that not only process manure into a useful gas that can be used for cooking, eating, and lighting but also produce a by-product from the process that is a char still rich in nutrients to return to the soil, which can be used for compost.

## Composting

There are a fair number of choices available to the composter who wants to explore the process, from apartment-dwelling worms in a homemade or store-bought wormery, to a giant pile out behind the barn. The composting projects in this book are obviously just a start, because once you have compost, you'll need to do something with it. There are often many local resources available through gardening clubs and organizations. We'll also discuss gardening projects a little more because, not only should you be worrying about what happens to your food once you've finished with it but also you should be worrying about where it comes from. The energy that goes into transporting food to our dinner plates can be enormous, and while there are often local farms and businesses you can choose to support, nothing is as close as your own back yard.

# Project 10: A Worm Compost Bin

Certain types of worms are able to digest large amounts of organic waste if kept in a suitable environment. It's possible to buy a complete wormery, or live worms, and have them delivered to your door—check Evilgeniusonline.com for more details—or you can build your own. Probably the most important thing to know about worm composting is that not everything can go into your composter. Meat and cheeses, for example, will tend to go bad before the worms can eat them, and start to smell something fierce. Eggshells and vegetables, on the other hand, are great worm food. There are complete lists and discussions all over the Internet about what you can feed worms as well as what doesn't work so well.

## You will need

- A plastic bin. A 20-liter bin will suffice for two people and a small number of worms. The size of the bin will depend on how many worms you plan to house, and how much food you'll be trying to dispose of.
- Two lids, one to place over your bin and the other underneath.
- A drill.
- Some rocks.
- Newspaper, cardboard, and other food scraps to start.
- A piece of carpet, a little larger than the top of your bin, or some other moisture-trapping layer (which still lets in air).

## Steps

1. You'll want to drill holes through the bottom of the bin and about a third of the way up the sides.
2. Fill the bottom of the bin with rocks, then layer some newspapers, a little soil, and some food scraps to start.
3. On top, place layers of moistened newspaper, covered with some damp cardboard.
4. Last, a layer to help keep the moisture in, such as a small square of carpet, fuzzy side down.
5. When your worms arrive (or you dig them up) bury them fairly deep in the pile and place a handful of food scraps on one side of the box. The worms will migrate to the food and leave behind their castings.
6. Feed the worms once or twice a week, rotating around the box each time. Some people try to feed them on one side for a week, then alternate; there are few hard and fast rules, so experiment!
7. When it is time, you'll want to harvest the worm castings (a pretty word for worm poop. Your plants will love it). One method is by feeding on only one side of the box for about 10 days to draw the worms over, then lifting up the cover on one side of the box to scoop out the castings.

# Project 11: Making a Bit of Methane



This project demonstrates the concept behind producing methane gas from your food scraps. Be aware while you are doing this that methane gas is very explosive, so take the appropriate precautions. We're not using the methane for anything, and though we won't have produced a lot of it, we will have created a little bit of a powerful greenhouse gas, which you'll want to get rid of safely. It is sufficient, for this small an amount, to release it outdoors away from any sources of open flame. Not terribly green of us for the moment, but by understanding a little more of how this works, I'm trusting that you will go out into the world and try capturing and using more of this valuable resource. If you are very, very concerned about this small amount of methane being released, you could put it in the middle of an empty field and throw a lit match at it from a distance. Better to plot a second system that incorporates use of the gas for your next attempt, than fret too much about it though.

Figure 4-1 *The materials you will need to make a bit of methane.*

## You will need

- Several plastic bottles, all the same size.
- Some plastic bags with mouths bigger than the top of the bottle, but not too big: the thin vegetable bags from the grocery store will work, as long as they don't have holes in them.
- Some tape.
- Food scraps (vegetable scraps ground in a blender would be ideal).
- A little bit of water.

## Steps

1. Cut the tops off of each of the bottles (see Figure 4-1).
2. Mix some food scraps and a little water together and put some in each of the bottles (see Figure 4-2). Each bottle will be a slightly different experiment, so it is best not to vary the amount of water or food scraps in each bottle for the moment (see Figure 4-3 for example).
3. If you are using a balloon, stretch the opening over the cut mouth of the bottle. Otherwise, place the bag over the bottle and tape it on so that most of the bag is hanging limp (see Figure 4-4).

Figure 4-2 *Adding water to some food scraps, once the top has been cut off a reused plastic bottle.*

Figure 4-3 *About equal amount of food scraps and water in two trials.*

Figure 4-5 *A warm place for the methane: at the exhaust vent of a refrigerator.*

4. Place your bottles in several locations. Try a cool place, or in the sun, or near another heat source (see Figure 4-5). Be creative!
5. Leave your experiment for several days.
6. Bottles in the warmer locations should have started to gather some gas in the bag. The balloon will inflate slightly. You've produced methane!
   Be careful when disposing of it, because it is explosive.

Figure 4-4 *Attaching the bag.*

## Larger systems

Larger-scale methane digestion systems are in operation around the world, primarily in hot countries because, as you've just found out, methane is produced best between fairly specific and warm temperatures.

### Online Resources

India and China have a large number of rural systems in place, and groups like Practical Action (formerly the Intermediate Technology Development Group—see www.itdg.org/) publish free booklets describing their operation in enormous detail based on years of experience. (Author's warning: the group's website is an addictive source of information on nearly every technology imaginable.)

Biogas systems (as they are called) are not trouble free, but they are relatively maintenance light as long as they are fed fresh ingredients regularly. If the microorganisms are kept in a stable and comfortable environment, small biogas systems processing human and animal manure and other wastes can provide a stable and low-cost source of renewable heating and lighting for families and villages.

## The rest of your trash

OK, so we've been pretty positive about the ability to extract energy from the organic waste that the average household produces. What about the rest of the waste? Because that trash is hauled away by truck, and processed and stored—each step requiring additional energy—it is wise to reduce your trash. One of the oldest mantras of environmentalism is: reduce, reuse, recycle. It is important to keep the processes in that order, and to actually pay attention to them throughout your daily life.

For example, the easiest place to reduce the amount of waste that is produced by your house isn't necessarily on trash day; it's on shopping day. The choices you make about which products come home in the grocery cart can have a large impact on the trash produced by your household. Most of the waste from your food products is a result of excess packaging, which is mostly unnecessary. Try to choose bulk foods over prepackaged ones, as they require less packaging (and thus energy) to get to the store, leave you with less waste, and usually cost less too. Some types of packaging are easier to deal with than others. You can also choose to leave the packaging behind at the store. You are buying a product, not more garbage. Making retailers responsible for disposing of product packaging is a most effective way of promoting less packaging in the first place. It is also efficient, so we talk about that more in the next chapter.

Your trash no longer contains your food and vegetable scraps (right?) so no need to worry about those. Separate your paper and cardboard if you haven't recycled it already; they can also be reused for gardening and in the compost. Glass jars can be useful to start germinating seeds and for storing fruits and jams you might make from the plants growing because of the nutritious compost.

Dangerous goods, including batteries, should be separated from your trash and taken to a hazardous waste disposal site. If you have old electronic equipment, try and find a good use for it by donating it to charity if it works. If it is broken and too expensive to repair, think about whether it was a useful expenditure of resources, and look into longer-lasting products next time. A life-cycle assessment, discussed in the next project, can help in this evaluation of larger items.

An obvious source of value in your trash is refundable containers if offered in your area. Make sure you pick these out and turn them in for the refund, it adds up quick. (Hint: remember this as a potential "Ant rule" for an upcoming project.)

What's left? A lot of plastic probably. Look on the undersides of the containers. There is usually a triangle of arrows with a number inside it, indicating the type of plastic. Based on this, you can figure out if you can recycle it in your area. Visit your local recycling station to find out what they accept. Your experience is likely to be that they do not accept much. Most of these plastics can release harmful toxins if burnt without pollution controls on the exhaust, so it is best not to do that.

## Recycling

It is likely that your community recycles cardboard, because the recycled product tends to have a value to paper producers, but not all types of plastics. Cardboard can in most cases be turned into more cardboard, while plastics, even when they are recycled, tend to be degraded by the recycling process. That means that when you return that 2-liter clear beverage bottle for example, it does not come back to you as another clear beverage bottle. Clear plastic bottles require a higher-quality resin than recycled bottles provide, so new resin is usually used. The recycled plastic will often go into a different product, such as plastic wood for outdoor benches, whose manufacture requires a lower grade of plastic resin. When possible, it is good to avoid packaging which when recycled loses value, though it can often be a funny trade-off.

Take glass for example, which can be almost entirely returned to more glass without losing much material value. However, glass is a heavy container. Depending on where your beverage was manufactured and transported from, it might be more energy efficient to get the plastic bottle, because of the amount of energy that can go into shipping glass over great distances. It's not always easy to consider these things when you are just trying to get a cold drink, but it is something to ponder for the longer term.

# Project 12: Return Your Trash

Briefly auditing your trash, as we've just described, often results in a pile of plastic that is not accepted by your local recycling depot. Too often this plastic is a result of overpackaging of goods, which probably didn't need it in the first place.

Your options for getting rid of it in a responsible way are limited and it might be easier to start thinking about why you have it in the first place? Many of the products we buy have a large amount of excess packaging, and it is in part of our responsibility as consumers to make it clear to those companies that we would rather not have to deal with disposing of it at the end. Dealing with that packaging costs us all money. If you do not have municipal trash services, you may already have to take it to a dump yourself. The taxes you pay go into ever-larger dumps, many of them filled with unnecessary plastics.

Legislation exists in some places that forces a company to take back all their packaging: you could try sending yours back. Another option is to make an effort not to buy overpackaged goods. An upcoming project goes into this in more detail. If it is something you very much want, write to the company and tell them why you are not buying their product. You might be surprised at how fast not spending your money can get results.

Often, letter writing is an effective way to get companies to consider the impact of their products and search out more responsible alternatives on their own, so you don't have to. But if you have no results, or would just like a faster response, this is a project for you.

## You will need

- Packaging: the sort that you didn't want in the first place.
- To know the store where you bought the item in the first place.

## Steps

1. Collect the excess packaging from items you recently purchased. They can be food, electronics, it is of no consequence, so long as you feel, having now thought about it a little, that the packaging was excessive or inappropriate or difficult to dispose of, or some other reason.
2. Return to the store where you purchased this item, with the packaging in hand.
3. Ask to speak to a manager or person in charge.
4. Give them the packaging and tell them politely what you are doing. Be firm that you are leaving the packaging at the store, because you do not feel that you or your municipality should bear the consequences of the packaging choices made. Encourage them to tell their superiors about the encounter, and to get back to you with a resolution.

## Small steps and giant leaps

One of the people who has often inspired me—as I've explored the subjects of the environment and life—started a blog a few years ago called the Ephemeral Tourist 2.0. Lately, it described the process he is taking to build-up the driveway of his house, as "Ant Rules." The premise is that ants can collectively do marvelous tasks by doing one of very few simple tasks it knows how to do, like look for food, or clear waste. Their "thoughts," as it were, have to be simple and have defined triggers, yet ants are capable of complex and amazing tasks working together. However, the number of "thoughts" an ant can have is so limited that it is incapable of surviving on its own. By contrast, humans could be considered "as the pinnacle of evolution. The creative potential of every individual human is marvelous. But the collective behavior of humanity is suicidal insanity."

The Ephemeral Tourist 2.0 considers, and I agree, that simple rules, like the ones we think ants follow, might be able to help us on our path to becoming better environmental citizens. One of his "ant rules," to improve the gravel content of the road crush (a blend of sand and gravel) that is his driveway, comprises the following:

1. Pick up a littered cup—like a McDonald's fountain drink cup, or a Slurpee cup—and place in [the] bottle cage of [his bike].
2. Fill with gravel-size rocks found in the street.
3. Place rocks on driveway.
4. Dispose of cup in proper receptacle.

As a result of doing this everyday on his commute to and from work, the gravel driveway is slowly growing. The Ephemeral Tourist 2.0 has found that

> Building it this way has me constantly thinking about value; about how we arrive at the purely subjective opinion that something has it, and how this ephemeral perception affects our experience of life. Within the confines of this small, four-step paradigm, I find that it imposes perceptions of value that didn't seem to exist before.

For example,

> a couple of rocks on the asphalt doesn't make much of a difference one way or the other. Maybe picking up a stone will keep it from being kicked up into a windshield by a passing car, or prevent a roller-blader from stumbling on it. But relocated to my driveway, along with all the other stones from my trips, now that rock has identifiable positive value. Further, in sum, open pit gravel mining destroys more of Alberta's landbase than tarsands mining. When complete, my hand-picked driveway will have made no net contribution to that environmental impact.
>
> suite-mck.livejournal.com; August 9, 2008
> reprinted with permission

# Project 13: Ant Rules

The objective of an ant rule is to make the attempt to see new value in something which either you or someone else has discarded. It proposes you create a "rule" for yourself that requires participating in revaluing that thing, whenever you come across it or are triggered into action. There are many examples throughout the book to help you think creatively about what can be an extended project. Try to provide a frequently reoccurring reminder that the value we ascribe to objects can change, and the label "waste" is subjective.

## You will need

- Patience and a desire to see the world in a new light.
- No fear of looking through trash is useful, but not necessary.
- To try to find a "rule" that consists of multiple small steps. Plot out the expected benefit you hope to get from it, and the environmental impact.
- Ideally, a frequently occurring trigger for your "rule."

## Steps

1. Look around you and think about how you can redefine or augment the perception of value in objects that surround you.
2. Start small. Picking up every refundable container and returning it to its appropriate space, as your first ant rule, is a good way to try this technique out.
3. Account for your actions and the impact (or avoided impact) those actions have. For example, we have discussed using discarded glass jars and compost to grow food. An ant rule could be that you will plant a new plant (and find a space for it) in every glass jar you would've otherwise thrown out. Your actions will have avoided the energy needed to recycle the glass (quite a few in a short time), as well as displaced transportation emissions from food that would've had to travel to you (if you grow edible plants).
4. Publicize your actions and be a positive example.
5. Your project, even if it takes a long time, not only serves a larger purpose but also serves as a positive and daily reminder of the positive changes you are in the process of making. Getting "green" is an ongoing process for all of us; no one has all the solutions yet. By thinking consciously about them and acting deliberately, we can make a conscious start.

# Underused land

In many cases it is possible to grow a much larger portion of the food we consume closer to us than we are currently doing. Oftentimes, an alternative is right there in the same store, and it is just up to us to look for it and make that part of our normal shopping routine. If you are fortunate enough to have some land to work with, it is of course possible to plant a large plot, such as in Figure 4-6. If you are not so endowed, there are also more proactive ways you can go about reclaiming underused space for more productive uses and in this way contribute directly to increasing the productivity of your local area and reducing the distance your food travels to you.

Particularly in urban areas, there is a dearth of underused land and space available for small gardens and individual plants. It is often a matter of looking at a piece of lawn a little differently and wondering what else could grow there. This upcoming project encourages that. Alternatively, it could be fruitful to look more closely at what has already started to grow naturally on disused pieces of land—by the side of the road, for example. Collecting native

**Figure 4-6** *Large garden plots, such as this one in rural New Brunswick that meets most of the vegetable needs of a family of four, are great for those who have the space.*

"weeds" is already a part of local tradition in many places. In Carelton County, New Brunswick, for instance, Fiddleheads—an immature type of fern classified as a weed—are a seasonal and local delicacy. There are local groups and sites, as well as books on the subject of what local varieties of plants are edible. You will likely want to talk to older residents who are also gardeners or farmers, as often the best knowledge of a particular area is only available orally.



## Project 14: Seed Bombs

"Guerilla gardening," a term dating to the 1970s apparently, accurately describes the act of planting productive plants on underused public spaces. The "movement" has spread across the globe, and there may be people in your area reclaiming underused space for productive uses. "Seed bombs" are a popular manner of quickly encouraging the growth of new plants.

### Tip

Non-native seeds can pose a risk to local species, so always make sure that you are not introducing a non-native species into your area.

### You will need

- Seeds native to your region (you'll have to check into it and it is *very important to check*), and ideally producing edible or useful material as well. Remember also that many of the plants we consider to be weeds are actually edible, but do look into it first. Some resources to help get you started are listed on Evilgeniusonline.com.
- Some compost and soil.
- Rubber gloves.
- Some vacant or underused land spotted nearby. (A required warning: it is illegal to plant something on someone else's property, even most public spaces, even if they are going to waste. Do be aware of this, budding geniuses, and don't get too evil. Perhaps your first spot of vacant land might be on a piece of property you own: something as small as a reused pot near a window is a start.)
- A mixing bowl and spoon.

### Steps

1. Mix your seeds with a small amount of compost and soil in the mixing bowl.
2. Add a small amount of water so that it will form into a ball and not fall apart.
3. Roll some balls, about the size of an egg. These are your "seed bombs." You may want to let them dry a little so they hold together, but not too much. It is ideal to do this pretty close to when you're ready to go out distributing your seed bombs (see Figure 4-7).
4. Ideally, dig a small hole with your spoon, and deposit the seed bomb inside. Cover it up with the soil you removed, and think about sprinkling some water on top as well.

Figure 4-7 *A completed "seed bomb" ready to be delivered to an unsuspecting plot of underused land.*

5. If you are using you seed bombs in a less than ideal location, you might just deposit it on some soft soil, and hope for the best.

6. Return in a while and check on your "intervention." Depending on what seeds you chose, you may have beautified your surroundings, or grown something that you could eat or use in another way. By producing it locally, you may have reduced the need to bring it in from somewhere else, whether that is a strawberry flown from California to New York, or one from the farm at the edge of the city you live in. Energy used to get that fuel into or to you has been reduced. There are lots of spaces to use more productively around us—it's up to us to use our genius to find them.

# Chapter 5

# Efficiency

## Heat sources

We've talked so far about different ways to conserve heat and cold, without mentioning where they might come from. There are a wide variety of heating and cooling sources that are in use in different buildings around the world. What type typically depends as much on the available resources and their cost, as on the overall efficiency of the system. Too often, when the person paying for the upfront costs is a different person than the person who will pay the running costs, inefficient heat sources are chosen. Electric heat has long been vilified in the environmental community for being an inefficient way to produce heat. We'll look a little more at how electricity is made later, but part of the problem is that a lot of our electricity comes from heat sources originally—like burning coal and gas in big power stations—which is then used to turn water into steam, then fed into a turbine which spins a generator to finally produce electricity.

Only about a third of the heat energy produced by burning the coal or gas becomes electricity, and too often the rest of that heat is wasted. Electric heaters can deliver almost a full unit of heat into a full unit of electricity, but the problem is that the electricity they are using has often already wasted two-thirds of the energy we had to begin with. Heat pumps (which we will also discuss in more depth later) can improve this situation. In many cases it is more efficient to burn the fuel directly where the heat is needed; however, this can create a host of other problems, such as local air pollution and poor indoor air quality if there is insufficient air circulation where the fuel is being burnt. Having a thousand little fires rather than one big one can also create efficiency and quality control issues, so the solution isn't always cut and dry. Add to this the problem of infrastructure. To move a given house or apartment with electric heat to another heat source often requires many modifications to the building in order to distribute the new heat source by air or water, as well as a method of venting the exhaust from the burning process.

In some areas of the world, district heating systems try to centralize the burning of fuels and then distribute the heat to other locations. These often make good use of heat from electricity production that would have otherwise gone to waste. When the heat is used over a larger area, these are called district-heating systems, while when both heat and power are produced and used on a smaller scale, like in your house, it is called a combined heat and power (or CHP) system. These systems can be very efficient because they produce two things that we use, heat and electricity, at the same time. Home-sized systems (see Figure 5-1) are coming onto

**Figure 5-1** *The Whispergen combined heat and power system is designed to integrate seamlessly as another appliance into your house. Photo courtesy of Whisper Tech Limited.*

the market, and we will talk about some of the technologies behind them a little later. These are all larger changes a person could think about making.

For most people in the immediate present, it is best to minimize energy use through conservation. Get aware of the inefficiencies around you, and plan to do something about eliminating those inefficiencies as fast as practically possible. Modern gas or oil furnaces are significantly more efficient than their predecessors, so think about upgrading if you can. Old wood stoves and open fires are now the largest cause of dioxins, which are carcinogenic and bioaccumulate in humans and other animals, so upgrading to a modern efficient pellet stove is also worth it. Heat pumps are getting more common for those with little alternative to electricity, but should always be combined with very energy-efficient building designs. Without any new fuel sources, just newly-insulated windows and walls, most buildings should already be seeing a drop in the amount of fuel in use, so congratulations. It's important to try not to spend that extra money on making the space warmer, but try to keep using it to make continual and gradual improvements to your space. If a plastic double-glazed window saves five dollars per month, for example, try saving it up for a winter season and having proper double-glazed windows installed by the next year.

## Light

As we've noted, the sun is one of the most readily available resources and often provides free heat, as long as we pay attention to capturing it. It also provides free light, if you hadn't noticed. Opening your curtains during the day and closing them at night achieves both in a simple way, but there are many more ways that more solar energy can be captured. If done carefully enough, there is enough solar energy to meet all our space heating and cooling needs. Indeed there is enough to meet all our energy needs, if we just use a little evil ingenuity. But there are always times when the sun isn't directly around to light our way.

Electric lights are clean, convenient, and efficient. Their use has been linked to a significant jump in human life expectancy, as they are often one of the first uses of electricity sought by those who have not previously had access to it. The alternatives, either darkness, or solid or liquid alternatives, release local air pollutants including particulate matter, which are carcinogenic, especially when used indoors. Particulate matter is often discussed in the media, in relation to smog. It comprises tiny particles suspended in the air around us and is often abbreviated as PM, followed by a measure of the size of the particles in subscript. For example, $PM_{10}$ is short for particulate matter of 10 microns (μm) in size, about 100 times smaller than a human hair. Particulate matter is bad for human health in general and contributes to changes in our climate.

### Tip

Where there's smoke there's particulate matter! Particulate matter is harmful to humans and other animals, but reducing it may also contribute to global warming. Particulates suspended in the air (like smog) are thought to reflect more heat back into space than clean air.

Incandescent electric light bulbs have been common for the better part of a century, and their economical production was a significant factor in the spread of electricity in the United States during the last century. Lately, however, it is becoming more common to find compact fluorescent (see Figure 5-2) bulbs on store shelves around the world. Some governments have even

**Figure 5-2** *An incandescent bulb and its package (right) and a compact fluorescent bulb with its package (left).*

gone to the somewhat extreme extent of banning the incandescent variety of light bulbs. What gives?

Well, the tried and tested incandescent-type light bulb has been implicated in doing something that a light bulb shouldn't really do: make heat. It's often more efficient to use heat directly than it is to make electricity (often from heat turning a turbine) and then use that electricity to make heat. This is especially true if the thing you are using to make heat, such as a light bulb, isn't designed to produce heat for our use. If you live in a part of the world where it is often very hot outside, and air conditioners are common, it is not hard to appreciate that having hundreds of little heat sources inside a space that is costing money to cool is not very efficient.

# Project 15: Installing an Energy-Efficient Light Bulb

We are letting the genius off a little easy with this one, but just for fun go check around and see just how many incandescent light bulbs are still out there. Look in your school or office, and remember those locations for the next project. There are lots of good statistics on how many power plants can be turned off if only one in five incandescent bulbs were changed to compact fluorescents. Rather than just telling you about it, we'll show how the calculations are made in the next project, so you can do your own energy accounting. The responsible evil genius should also consider that it is not efficient to throw out a perfectly good light bulb, but instead wait till it burns out, then replace it with an efficient one. Also, these new compact fluorescent bulbs have their own problems, something we're only starting to become aware of.

Fluorescent bulbs use mercury in them, which is a very toxic and poisonous metal. This is the same substance found in mercury thermometers, and even a small amount is quite dangerous to humans and other animals. This means that these bulbs should not be thrown into landfills, because the mercury can be released and seep into undesirable places. It is therefore very important to take spent compact fluorescent bulbs to a hazardous waste disposal location or a recycling system that is ready to accept them. Because this is still a new technology, many recycling depots are not ready, so ask questions. The hazard caused by mercury in fluorescent bulbs is the danger of them breaking, which would release the mercury into the local area. A little research on the types of precautions you can take in this event will go a long way, but ultimately it will depend on the place it happens. You will in any case want to remove yourself from the room in which a bulb broke, and open some windows; then, vacuum the area where the bulb broke and collect the broken shards with gloved hands. Seal the vacuum bag in another airtight bag before disposing of it in a location that can contain the mercury. This is not pretty stuff, so be careful; the best thing to do is not break the bulbs in the first place. There are special high-impact compact fluorescent bulbs available that are stronger and won't break as easily for locations where they might be at risk of doing so.

## You will need

- A compact fluorescent light bulb (see Figure 5-2).
- A working light socket in which the old incandescent bulb has burnt out.
- A good wrist.
- Possibly a ladder.

## Steps

1. Get you ladder set up if needed.
2. Unscrew the old bulb by turning counterclockwise.
3. Screw in new bulb, turning clockwise.
4. Turn on the light.
5. Get busy calculating your energy savings.

### Definition

What is a watt? A watt is a measure of the rate of energy use, and is not confined to measuring electricity. A person riding a bike up a moderate hill does work at a rate of about 200 watts (W). If they rode up that hill for 2 hours, we could express that as 400 watt-hours (Wh).

# Project 16: Calculating Your Energy Savings from Project 15

This is the most amazing part of a very simple project, and something you can repeat and adapt to almost every project in this book. Try going forward with these basic ideas and apply them to the rest of your energy consumption. If you are intrigued by the idea that little actions can save a lot of money, go hunting for an energy meter (see Figures 5-3 and 5-4), which is an extra tool in this project but useful for the next one, too. Some municipalities distribute them for free, because it is a valuable tool to get residents aware of their power consumption habits. Other times they can be found at your local hardware store, or in energy-efficiency sites listed on Evilgeniusonline.com. Permanent systems that display the current rate of electricity use in a convenient location, like by the door, are coming on to the market. These permanent systems have been shown in studies to dramatically improve the likelihood that house occupants get aware and reduce their electricity use by turning appliances off before leaving.

### Definition

A kilowatt-hour (1000 watts for 1 hour; abbreviated kWh) is a measure commonly used by electricity companies to bill residential customers for their power consumption. 1 kWh is, for example, consumed by 10 100 W light bulbs burning for 1 hour, or 10 25 W bulbs burning for 4 hours.

## You will need

- A pen and paper, possibly a calculator.
- An electricity consumption meter (see Figures 5-3 and 5-4).
- A corded light.
- Your old light bulb, and the new energy-efficient one (if using the meter).

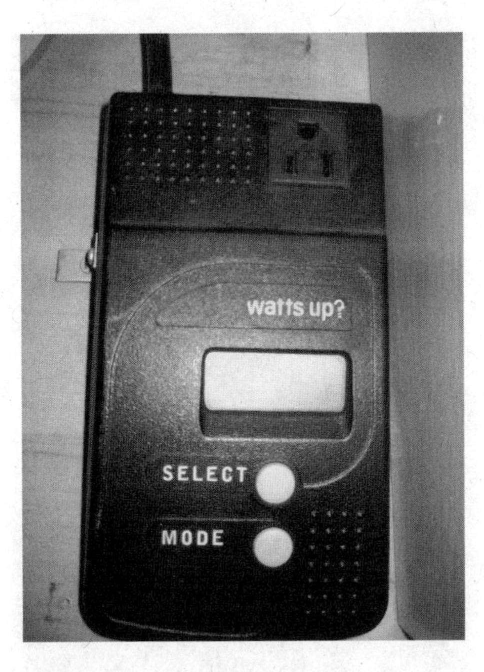


Figure 5-3 *A branded energy meter.*

## Steps

1. Copy down and fill out the following table using the values you find on the light bulbs themselves:

   Energy consumption of old light bulb:____ watts (A)

   Energy consumption of new light bulb: ____ watts (B)

   Number of hours used per day: ____ hours (C)

Figure 5-4 *A generic UPM energy meter.*

2. On your piece of paper (or calculator, if needed): Multiply the energy consumption of the old bulb by the hours used (A × C) to obtain your old watt-hour consumption figure (call this D). Then multiply the energy consumption of the new bulb by your typical usage (B × C) and call this E.
3. Now subtract E from D, and you have the energy savings in watt-hours. Shown another way: (A × C) – (B × C) = ____ Energy savings in Wh
4. To turn this number into something that is meaningful, like money, have a look at your current electricity bill for a figure that shows the amount they charge you for a kWh of electricity. A typical value in the range of 5–10¢/kWh is normal in North America. United States residents can go to www.eia.doe.gov/fuelelectric.html for a profile of their state. Multiply that figure by your energy savings to estimate your savings.

## Variations

5. Using the energy meter, plug the light with the old bulb into the meter and then into the wall. Turn the light on and record its consumption on the digital display. Most of you will note that the light consumes electricity at a faster rate when starting up, so you can leave it on for a little while and use the steady-state value. Repeat for the new bulb, and input your values into the equation above.
6. A more detailed meter will actually allow you to plug in the light for an hour, and estimate how much money that costs based on average electricity rates. Try cycling through the options on the meter.
7. If you are in a location where you would like to demonstrate the difference to a variety of people, a display like the one shown in Figure 5-5 is an excellent learning tool. An alternative, which employs muscle power to convey the impact of the extra work required to light an inefficient light bulb is shown in Figure 5-6. An old exercise bike is attached to a 12 V generator, shown in detail in Figure 5-7, which powers one of two bulbs the rider can switch between. It is immediately obvious when pedaling, which type of bulb needs more work done to provide the same light.

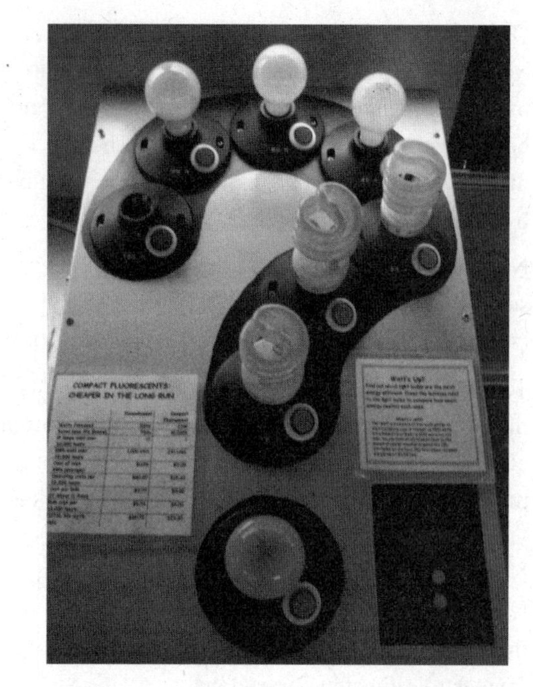

**Figure 5-5** *A demonstration unit that shows the energy consumption of a variety of bulbs, and allows users to notice that the same amount of light can have large variations in energy use. Shown on Big Blue, the Climate Change Awareness bus run by the Falls Brook Centre.*

**Figure 5-6** *A tactile demonstration of the amount of work needed to light a bulb, by pedaling an old exercise bike attached to an electric generator. The generator powers one of two lights in front of the rider. When the switch is flipped from the incandescent to compact fluorescent, the work the rider must do to light the bulb is much less.*

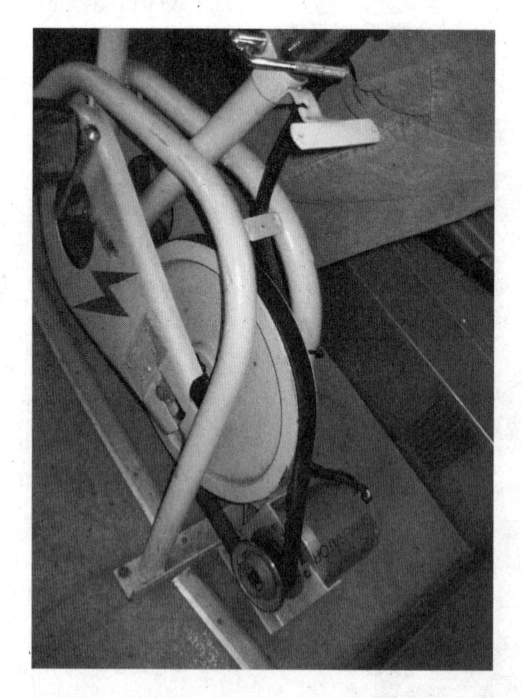

**Figure 5-7** *Details of how the exercise bike is attached to the generator, for you truly evil geniuses who will undoubtedly try this at home.*

# Project 17: Auditing Your Appliances with an Energy Meter

If you went out and got an energy meter and have it handy, why stop with the lights? Pull out your electronics and see just how many devices around the house are using electricity, often when you're not even looking!

## You will need

- A Watt-Wise meter or something similar that measures the energy consumption through its provided outlet.
- Some electronics. To start, find a TV with a standby function (usually those with a remote control operate on standby functioning).

## Steps

1. Plug your TV into the Watt-Wise meter. Leave the TV off.
2. Look at the number of watts used by the TV when it is off.
3. Turn it on if you are curious, but notice that some electricity is used when the TV is off. Go find another TV if that is not the case, or try your DVD player or stereo amplifier.
4. A typical TV consumes 4 W when it is turned off. Multiply that number by 10 hours, a typical length of time to leave a TV off waiting to be watched.
5. Recognize that an astonishing 40 Wh used by the TV when it is off is more than is used by the energy-efficient light bulb we installed earlier (typically 23 W) used for an hour a day.

## Ghost loads

An enormous number of appliances and electronics use power when they are not on or being used. Some of them are useful, like the wireless router that allows you to connect to the Internet without having to remember to plug it in, but others are less so. Home entertainment systems and many other electronics around your home are often fairly poor performers when it comes to using power while turned off. This is often referred to as a ghost load. If you use the energy consumption figures from the last project and project them for the life of some appliances (microwave clocks are notorious) you might find that the ghost load (when it is waiting for you in standby) of your appliance uses more energy over its lifetime than when it is actually doing something useful for you. Astonishing, isn't it?

In the UK and many other parts of the world, people are accustomed to power outlets that have a small switch attached to them, which allows the power to be switched off while something is left plugged in. This is useful for energy-conscious consumers in those countries who have to "just" remember to turn off the switches around the house before they leave, or when they are not using something in particular, like the stereo. I have heard that the reason for these switches is that the 220 V power supplied to the outlets in the UK is a little more likely to kill a person if they were to get accidentally electrocuted while plugging or unplugging devices, so the switches are a safety precaution. For us in North America without these handy switches, minimizing ghost loads is a little more difficult, but certainly possible and just as important.

# Project 18: Getting Used to "Better Than Off"

In the UK and other parts of the world, electricians tend to install switches at the outlets that allow a person to turn off appliances that tend to have ghost loads. This leaves a person in those parts of the world the not-inconsequential challenge of remembering to turn off those switches. Those of us in North America have an extra challenge in dealing with ghost loads directly. We could go around unplugging everything that is not being used, but some people find that inconvenient. So our evil genius will illuminate a third way. On a long-term basis, this should illuminate a fourth step, less dependent on the fragile and fallible human memory.

## You will need

1. Power bars with switches on them, enough to plug all your electronics into, like the one pictured in Figure 5-8.
2. Reminder stickers or sticky paper and bright markers or a printer.

Figure 5-8 *A power bar in use. Note the power button to the right of the plugs, which is not normally perceived as an energy conservation tool, but if used correctly could save you money by eliminating "ghost loads."*

## Steps

1. Unplug your TV, stereo, DVD player, and other appliances from their wall socket, and plug them instead into the power bar.
2. Plug the power bar into the wall.
3. Turn on the switch to watch TV.
4. When you are done, turn off the switch to get rid of ghost loads.
5. (Optional) Modify the table in Project 16 to calculate how much energy you are saving by turning off your electronics when they are not being used, and how long those power bars will take to be paid back.
6. If you are like me, you might forget to turn off those handy power bars we just installed. A solution is to make helpful "remember to turn this off" stickers and place them around your house. You might put a sticker around the light socket to remind people to turn off the lights, and another one by the door to remind you to turn off the power bar near the stereo before you leave the house for the day. Be creative!
7. A longer-term solution is to look into systems that can turn outlets on and off from a central location (like a main switch near the front door), or ones which can do so automatically based on your schedule.

## Auditing

Auditing your energy consumption is not very difficult. In fact, we started the process in earlier projects. An audit can be as simple as taking account of all the uses of energy in our lives, usually (but not always) with a conservation purpose in mind. In our case the purpose is to reduce our energy consumption so that the audit should reveal places where our energy consumption is particularly high. There are a number of tools available to the home energy auditor, especially those downloadable from the Internet (see Evilgeniusonline.com), but it is also possible to use some math to get a good approximation. One common type of online energy-conscious audit calculator goes by the name of a carbon footprint calculator, which converts energy consumption from a variety of common sources into a carbon equivalent. Many of these calculators make the point that North American consumers tend to use an average of five times as much energy, and consequently contribute that much more carbon, as most other citizens of the planet.

To start your own audit, whether on paper or online, start by going around your house and gathering the energy consumption figures for all the major appliances in your house. Then get a hold of your latest utility bills, provide fair estimations of about how much fuel you put into your car every week, and how often you take a national or international flight. For most Americans, transportation is almost half of their individual energy consumption, so small steps here can have a large overall effect. More projects are covered later in the book to help you cut your transportation pollution, but the very first step is to be conscious of the fact that every time you decide to drive rather than walk, or fly rather than take a train, you are contributing more carbon to the atmosphere than you have to. This is because driving uses fuel and therefore emits carbon into the atmosphere, while walking certainly does not, and a train tends to emit fewer tons of carbon per kilometer of passenger travel than an airplane. Even better is finding ways not to take unnecessary trips, by using video conferencing and other information and communications technologies more effectively.

## Food miles

One aspect that will become apparent in your audit, where transportation contributes significantly to the energy we consume, is the distances that commodities have to travel to get to us. For goods we use on a regular basis, like food, which can be grown in many places, it can be astonishing to realize how far they now travel to reach our grocery store. We are privileged to be living in an era of high-speed air travel when Mexican bananas can reach supermarkets around the world before spoiling. However, it becomes an interesting question for an environmentally concerned genius to consider how much work went into growing and transporting the fruit, compared to how much work they would do after eating it. There are of course many other factors to consider when choosing your food from the grocery store, including how much enjoyment you (as an individual) get from eating particular types of food.

Some types of food that many people enjoy can only grow efficiently in certain climates, whereas others have a greater tolerance. Greenhouses can help extend the growing season, if heated throughout the year in many climates, but there is an energy cost involved. There are also certain perverse factors at play in the pricing of some foodstuffs, which I have never managed to fully understand. For example, I found that in the mountains of Wales, where sheep abound, New Zealand lamb was less expensive than its Welsh counterpart in the grocery stores. I understand New Zealand doesn't subsidize its farmers, and so I struggle to explain why such a large distance traveled doesn't increase the price more. It seems obvious that there would be a higher amount of energy consumed by eating meat from halfway across the world, but it is not always a simple matter.

There are also, in UK supermarkets, both locally grown tomatoes and tomatoes imported from Spain. Locally grown tomatoes are available year-round but, because of the British climate, they are grown inside heated greenhouses. In both cases energy went into producing the fruit, and some have estimated that less energy went into the imported tomatoes, because more energy went into heating the local greenhouse than was used to transport outdoor grown produce from abroad. It is sometimes a close call when large energy consumers like heated greenhouses are involved, and there are often no definitive answers, but it is important to open the debate and consider the implications of our consumption habits.

## Life cycle assessment

It is possible to further broaden the scope of the assessment of your consumption patterns to account for the full life cycle of a product. For example, one level of audit or assessment of the light bulb from the earlier project would be to compare your monetary costs, as we began to do. Add up the cost of the bulbs, the estimated number of hours of light they will provide, and how much the electricity to power the bulb will cost. An incandescent bulb will cost less to purchase, but will also provide fewer hours of light and cost more in electricity, and over its lifetime will cost more money. Looking further into both bulbs, a life cycle assessment (LCA) could look into the impact of disposing of both products, including recycling and containing the mercury in the compact fluorescent bulb, as well as where the resources to produce the bulbs and where the electricity that powered the bulbs came from.

The LCA is a powerful tool to illuminate the cradle-to-grave impact of a variety of products and commonly employed by industry. Developed in the 1970s as an accounting practice for energy consumption, the LCA can quickly illuminate negative side effects in seemingly benign developments. However, LCAs must be put into perspective and balanced against the real joys and pleasures that we gain from some actions and products. For example, strictly limiting your personal food miles can severely restrict your diet. When I was first exposed to the notion that my eating habits had a large transportation energy impact, I reacted by dramatically curtailing my consumption of foods I truly enjoyed. Thankfully, I was not alone in having had this sort of pang of eco-guilt. "How I got over my Eco-Guilt and learned to love the banana: Lifecycle assessments reveal the true cost of products we use" has been reposted at the Ephemeral Tourist 2.0 (from the last chapter) at my request. The avid reader who may be on the verge of never again tasting a fresh guava is urged to read it.

# Project 19: A Responsible Shopping List

A responsible shopping list can take many forms. It could be criteria you note down or have in your mind, by which you judge products on the spot in the store. It could equally be an actual list of products you either will or will not buy on the basis of the performance of the product or the company in meeting your standards, which you have researched beforehand. With the growth of mobile Internet access, there is also the potential do some research while you are shopping, and some interesting projects are trying to make that easier. Like the dreamers at one of my favorite sites, halfbakery.com, students at UC Berkeley are hoping to bring a mobile Responsible Shoppers Guide to the public, which scans barcodes and reports a score for the product based on ethical, social, and environmental criteria.

### Online Resources

A responsible shopping guide idea, filled with useful resources on barcodes, and some links to responsible shopping websites: www.halfbakery.com/idea/X-ray_20specs_20for_20consumer_20products

The ibuyright program in development by UC Berkeley: www.ischool.berkeley.edu/programs/masters/projects/2006/ibuyright

Co-op America's Responsible Shoppers Guide has many resources and is a good place to start looking into the effects of what you buy and alternatives: www.coopamerica.org/programs/responsibleshopper/

## You will need

- A list of products you commonly buy or major purchases you are considering making.
- Some time to do research and digging, not only into the items you normally buy but also into their alternatives.
- The patience to try new things and upset your routine as well as the wisdom to realize that for some things, right now, there may not be an alternative that is suitable.

## Steps

1. Start by looking into the products you normally buy. If you are accustomed to the products and have been a loyal customer who now finds something they are uncomfortable with, it is worth your while to communicate directly to the company. Be clear and concise about your issues and explain that you have been a loyal customer who has new concerns that may affect your future purchases.
2. If you find the company making the product inadequately addresses your concerns, start looking for options. Online resources are a starting point, but developing your own knowledge base on which to judge is invaluable, so do a lot of reading.
3. Once you have made a list of items or criteria for your purchases, refer to it and work at incorporating those values into small and large purchases. When the time comes to buy a new TV, for example, look for one that has no "ghost-load" when its off.
4. For those items where there isn't an alternative, or where an alternative does not perform to the same standard, be judicious in its use. Continue to communicate with the manufacturers of both your current product and the potential alternatives. Either one might soon come up with something better, especially if a customer or potential customer asks.

# Chapter 6

# Using Water

## Introduction to water

Water is essential to life, and has a great many uses that are both essential and non-essential (see Figure 6-1). We cook with water, bathe in water, and use water to transport away wastes we'd rather not deal with (like in our toilet). It is useful to heat water as it stores heat well. We use the properties of water to cool ourselves, as well as keep things we would like cooled cold as ice. We can also store and move energy around using water, which we'll look at later.

Providing clean water through pipes right to our house uses energy, different amounts depending on where that water comes from and how scarce a resource it is. In areas where there are bodies of water nearby, cleaning could occur through a variety of ways. If a city is close to the ocean without large freshwater sources nearby, it might require desalination plants, which may require large amounts of energy. One method of desalination is to heat the water to the point of evaporation, and to collect the evaporated water, leaving the salt behind.

Where a water source is not salty, it is often sufficient to heat water enough to kill any harmful bacteria or pathogens. In a large plant supplying water to many households it is probably economical to have a laboratory of some sort to test the purity of the water before it goes into the pipe network. If you are in a more remote location or somewhere else where the quality of the water supply isn't likely to be tested properly, there are still simple things you can do to improve water quality. This is a common problem in parts of the world where incomes are scarce, and some ingenious people have come up with several ideas and inventions to help provide clean drinking water using the sun's energy. Of course the sun can help us with boiling water, but that often times requires concentrating the sun's energy in some way. Pasteurizing water is another option, which involves raising the water temperature to less than boiling, which is easier to achieve using solar power without concentrating devices. Heating water directly with the sun's ultraviolet (UV) rays does something else useful to water: it kills organisms in it that are harmful to humans and makes pasteurization a safer, low-cost option for many situations.

**Figure 6-1** *A water fountain in a hot city uses the evaporative cooling effect of water to moderate the temperature in a park. Photo courtesy of Valsa Shah.*

# Project 20: A Simple Water Purifier

Pasteurization is a process that uses heat to destroy organisms in water that are harmful to humans. It uses less energy than boiling water for 10 minutes, which would result in sterilization of the water and the death of all organisms in the water. The difficulty of pasteurization is in knowing that the water has reached a sufficient temperature for all harmful organisms to be killed. You could use a thermometer to test the temperature of the water or, if you were going to do this a lot, a pasteurization indicator, which, when placed inside a simple solar pasteurizer, indicates that the water is safe for human consumption. Solar pasteurization adds one more layer of safety to the process, by exposing the water to UVA radiation (ultraviolet radiation of 320–420 nm), a particularly beneficial sort of radiation for destroying harmful organisms.

## You will need

- At least two small plastic bottles (the recyclable kind) often used for water or soft drinks. Clear plastic ones work best, and various sizes could work, depending on your needs, so it would be good to have variety for experiments.
- Some black paint and a brush.
- A thermometer suitable for taking the temperature of water. An inexpensive digital oven thermometer works well in this role for those experimenting (see Figure 6-2).
- Alternatively, for those with a longer-term interest in this process, a water pasteurization indicator (WAPI) can be purchased for a minimal cost from Solar Cooker International or made with their directions (at solarcooking.wikia.com/wiki/Water_pasteurization) using the following materials:
  - Fishing line.
  - Stainless steel washers.
  - A thin plastic (or PVC) tube (about 1/4 inch) with end plugs or a method to seal the ends (such as melting the plastic together with a propane torch and pressing the tube flat together with pliers).
  - Some wax that has an appropriate melting point. This can be tricky, as the wax must melt at the correct temperature for the fluid that is being pasteurized, in this case water (so about 150° F (65° C)). The world experts in this field recommend Myverol 18-06 K wax. You may have to do some digging for this, or experiment with homemade waxes and a thermometer.

## Steps

1. We are going to test two pasteurizers side by side, simultaneously exploring water pasteurization and the power of black to help catch more of the sun's heat.
2. Clean your recycled plastic bottles and remove the label.
3. Paint half (dividing the bottle vertically) of one bottle with black paint and let it dry (see Figure 6-3).
4. Fill two bottles with water, take the temperature of the water (see Figure 6-2), and then put the caps on them.
5. Place the two bottles in the sun. Put the black side down for the painted bottle.
6. Return after several hours (6 hours or more), and check the temperature of both bottles. In order to be considered pasteurized, they should have reached at least 65° C (150° F) for a period of time. If they have reached that temperature using direct sunlight alone, they have also received enough UVA rays to be an effective guard against water-borne disease and infection. The WAPI can tell you automatically if the water has reached that point, otherwise you can keep using a thermometer periodically.

Figure 6-2 *Taking the starting temperature, with an inexpensive digital oven thermometer, of the water inside one of two large solar water purifiers. A third smaller version is also visible on the left.*

Figure 6-4 *Two large solar water pasteurizers in the sun on the right, one black and one clear. To the left is a smaller version, then the solar cooker from the next chapter with a fourth pasteurizer inside.*

7. Depending on where you are in the world and how much sun is available, 6 hours may not be enough to raise the temperature sufficiently. Check to see if there is any difference between the painted and clear bottles: you might be surprised! If neither is warm enough, don't worry about moving to a warmer climate yet. Keep reading and we will reuse our water canisters in later projects with some solar reflection, concentration, and a bit of insulation to raise the heat a bit more (Figure 6-4).

Figure 6-3 *Two reused pop (soda) bottles, one has half of it painted with black paint to improve the rate of solar absorption.*

## Cool water

When the sun is out, it can often get too hot for most of us to feel comfortable. One of the things that helps me feel better on a hot day is a cold drink; unfortunately, there are times when there isn't a cold fridge nearby, or if there is, maybe there is not electricity to power it.

### Tip

Most people have experienced evaporative cooling but probably not noticed that was what was going on. On a hot day, pour some water over you head and feel yourself get colder. Evaporative cooling operates when water evaporates and takes some of the heat surrounding it with it, leaving what one of my professors called "coolth."

The idea for this project comes out of a trip I once took to the desert areas of Kenya. The town had once been prosperous owing to a successful lodge in the area that had it's own airstrip and electricity provided at the lodge and into the village. When the lodge owner moved back to Europe and there were no more foreign visitors, the electricity became sporadic and then one day stopped altogether. People in the village who had bought electronics had no use for them anymore, and those left with refrigerators they had bought to cool drinks for their European visitors, were left, presumably, without cold drinks.

Except that I got served a nice cool drink on a day that was 40° C (104° F) in the shade, by a man in a tin shack that had no electricity. He pulled it out from a fridge that was turned on its back, so the compressor was on the ground and the door opened up. He had built a second box, smaller than the refrigerator compartment and placed it inside the fridge, then filled the space in between the walls of the fridge and the box with water and some sand. The new and improved fridge worked without electricity on the principle of evaporative cooling, and worked pretty well on that 40° C (104° F) day. This easy but useful project is another way to demonstrate the same principle – I think I've given enough details for you to convert an old fridge already! (See Figure 6-5 if not.)

**Figure 6-5** *An experimental, modified Zeer fridge made from common household mixing bowls. The vegetables inside stayed fresh until used.*

### Tip

This sort of system is also known as a pot-in-pot or a *zeer* in Arabic. They are gaining in popularity in hot areas of the world with the advent of portable versions that allow food to be kept fresh longer without using any energy. H2G2 has an introductory article here: www.bbc.co.uk/dna/h2g2/A2116766.

## Project 21: Evaporative Cooling

The materials and directions are purposefully vague in this project, to enable real evil geniuses to make adjustments for their own situation and to scavenge their own scrap materials. Reusing materials outside, or after their original use, is among the very first tenets of a responsible environmental lifestyle, easily remembered as the 3Rs: reduce, reuse, recycle (yes, in that order, it's important).

## You will need

- A can of pop (soda) on a hot day.
- A wire cage bigger than the pop can (make your own out of coat hangers or other materials: a similar project calls for a bamboo basket). It is important that it is roughly a tube shape, and big enough to hold the material in shape around the can without touching the can.
- An old sock that is longer than the cage and can, or some loose fabric and a needle and thread (good to have handy).
- Some water, preferably in a spray bottle.
- An old plate or other base to put the can and cage on.
- Some pieces of a broken plate or flat stones, or other alternatives, which together should be smaller than the plate, but bigger than the can.

**Figure 6-6** *Salvaged metal pieces form a base for the bottle of pop on top of a water-filled plate. A metal cage covered in damp fabric is being lowered over the top to induce evaporative cooling of the drinks.*

## Steps

1. Fit the sock around the wire cage.
2. Place pieces of broken plates as a base on top of the unbroken plate.
3. Place the can of pop on top of the broken pieces of plate or other base (see Figure 6-6), and the wire cage with sock over the top (see Figure 6-7).
4. Fill the plate with a little bit of water and make sure the sock or fabric hangs down into the water.
5. Cut a small piece of sock or other fabric as a loose-fitting lid, or try just draping the fabric over the top (see Figure 6-8).
6. Spray the sock down with water until it is moist, not damp.
7. Leave outdoors in the heat (not necessarily the sun).

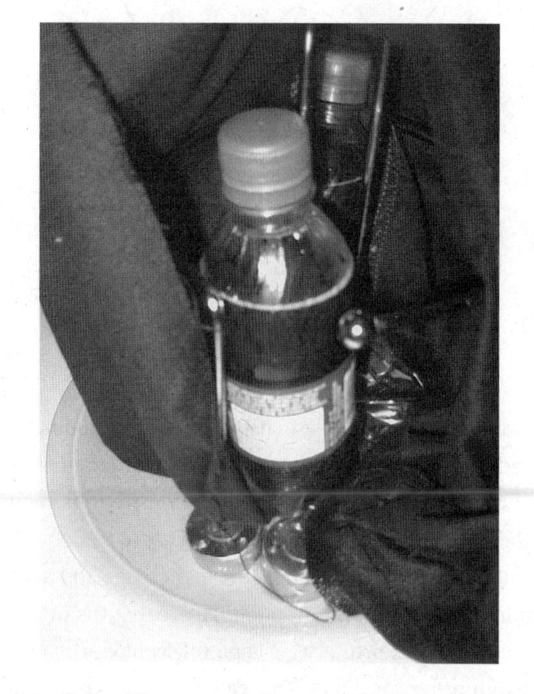

**Figure 6-7** *The cage of the easy-to-build evaporative cooler is shown.*

Figure 6-8 *A simple evaporative cooler for two small pop bottles in use.*

8. Spray occasionally as needed to keep the cloth damp (the water is evaporating—this will make your drink cold).
9. Remove the caged sock after a few hours, and enjoy a cool drink!

## Water efficiency

It is important to conserve water, whatever its source, because it is a valuable resource. So valuable is this resource, that some people predict that water will be the reason for future wars and conflicts. Energy goes into cleaning and purifying water, no matter what its source, and oftentimes in populated areas energy goes into cleaning water that has been soiled by human uses. We saw in the last project how a degree of clean water can be achieved without enormous energy use, but that is not usually the case.

# Project 22: Saving Water Around the House

As we discussed in the last chapter, many environmentally friendly behaviors involve a technology component as well as a human element. Visual demonstrations such as the one pictured in Figure 6-9 can serve to help explain the ability of simple new technology to have a dramatic effect on consumption. However, it is often up to us to remember to also change our behavior in simple ways, in order for many new technologies to have all of their intended effect. In the case of water, little actions—like remembering to turn off the water while you brush your teeth or lather up in the shower—can dramatically lower the amount of water your house consumes. It sure does seem silly to spend a lot of energy and resources getting clean water all the way to your house, only for you to let it run straight back as dirty water, when it is still pretty clean. It's just a waste. Again, there are a lot of different organizations around the United States and the world that have produced helpful stickers and reminder notices that you can put up around your house and office to help you remember to turn off the water when you are not using it. There are lots of other places to look for water waste around the house: if you have an automatic sprinkler system, for example, think about adjusting it manually when it is raining outside because nature has already done the work for you!

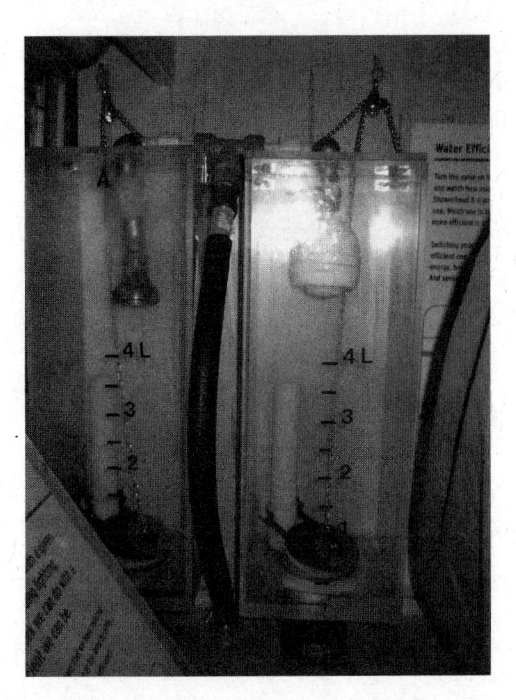


**Figure 6-9** *A dramatic demonstration of how much less water is used by a low-flow showerhead, on the Climate Change Bus at the Falls Brook Centre, New Brunswick. Both showerheads empty into graduated sealed containers using the same water pressure. The message: for the same shower, a low-flow showerhead uses less water.*

## You will need

- A low-flow showerhead (found at most hardware stores nowadays or given away by municipalities and water providers—read the Steps before calling them though).
- A toilet bag (also given away by municipalities, sold at hardware stores, or created yourself from a sealed bag and a clip).
- An adjustable wrench.
- Some silicon thread tape (also at the hardware store).
- More stickers or tape, paper, and markers.

## Steps

1. A low-flow showerhead replaces your existing showerhead. You may already be using one without knowing it, so check first. One give-away sign is a little metal cylinder sticking out a little, just between the thread and nozzle (see Figure 6-10). This cuts the flow of water completely, so the very concerned genius can switch the water off while lathering up, without having to readjust the temperature to rinse.
   - Make sure the water is off, then use your wrench to unscrew the existing showerhead. Peel off any old silicon tape; then add a layer of fresh tape to prevent leaks, and screw on the new low-flow showerhead.
2. A toilet bag displaces a percentage of the water that is stored in the tank of your toilet, waiting to be released for a flush. Ordinary toilets can use up to 20 liters (5 gallons) of water per flush, while a low-flow toilet can use as few as 2 liters. A toilet bag simply reduces the amount of water that is used in each flush, by displacing it with water kept in a bag through many flushes.

**Figure 6-10** *A low-flow showerhead installed. The on–off control is arrowed.*

- Fill up the bag with water from the sink. If there is a clip, use it to attach the bag to the edge of the toilet. The clip typically holds the bag out of the way. An alternative is simply to put a capped 2-liter bottle filled with water in a location in the tank where it will not interfere with the flush mechanism.

3. Contact your municipality or water provider and ask them about having a water meter installed, if you don't have one. Unmetered water is still common and contributes to wastage. Methods similar to the auditing process we have done for electricity can be applied to water as well.



# Project 23: Collecting Rainwater

The water that reaches the taps in your home is energy laden, so you've made every effort to conserve it in the last project. Along the way, you might have wondered if you really needed the very clean water your water provider puts in the pipes, for everything you use water for. After all, we use water for a lot of things. If you live in an area where water is metered and charged for, you might have noticed that there is also an awful lot of it that literally just falls from the sky. It's only common sense to use something that is free if its just as good: right? For some purposes, rainwater is just as good. The system pictured in Figure 6-11 is used for watering the garden and washing up in a workshop, two excellent uses for rainwater.

Collecting rainwater can be as easy as giving the water somewhere to go and stay for a while. Often the roofs of houses have gutters and spouts, which typically deposit this water onto the ground, where it runs off to the drains. By collecting the water above ground level, you can let gravity work to provide some pressure in the pipe, to deliver the water where it is needed. If you have a roof to collect water from, you'll just need a rain barrel. They are common in hardware stores, or if you're keen

**Figure 6-11** *A rainwater collection barrel mounted on the back of a workshop at the Falls Brook Centre, New Brunswick. The soak below the barrel catches the oily particles from the biodiesel workshop's gray water waste.*

**Figure 6-12** *View of the rainwater barrel (left) outside and sink (right) inside the workshop. The bottom of the barrel is only slightly above the sink tap, but provides more than enough pressure for washing up.*

**Figure 6-13** *This inexpensive commercially sold weather system can assist in understanding the rainwater resource available in the local area and is helpful when planning for a larger system.*

to find a reused one, olives are often shipped in containers about the size of the barrel in Figure 6-12 and a larger grocery store or restaurant may have a few around. Once you've collected the water, you'll have to think of a way to get it out of the barrel: either a pump or a spout drilled into the bottom (if its elevated) will do the trick.

If you do not have a roof to collect water from, then this project is designed to give you ideas. Usefully, it will also give you a simple way to measure the rainfall over a period of time, so you can make a better estimate of the size of container you will need for a larger system. The small collector pictured in Figure 6-13 would be not too different than what we used to take a spare room off-grid: compare the average resource (rainfall × surface area for water to be collected from) to your needs (number of liters required daily × days of independence required in between rainfalls).

## You will need

- A (loosely) graduated container. A 2-liter (half-gallon or smaller if you live in a dry area) pop bottle will do. Take a marker and make eight equidistant marks up the side.
- A funnel or materials for its construction from:
  - stiff cardboard, preferably plasticized or waterproofed—a US letter-sized piece will likely be enough for a 2-liter bottle, depending on where you live
  - if you can find no waterproof paper, then a large garbage bag
  - and some tape
- A few rocks or a stand to support the bottle and funnel upright.

## Steps

1. To construct your own funnel, make the two pieces of cardboard into individual cones by pulling corners along one long side together and over one another. Then slide one cone into the other, so the large ends spread out. Make sure there is a hole in the bottom of your funnel! If not, then poke one through the bottom.
2. Use tape to fasten your funnel together, and fasten it to the top of your collection vessel.
3. Stand it upright, and secure it so it won't fall down. It is certainly top heavy at the moment, but remember that a little water in the bottom can help it stay upright. Just make sure you note what you added before, so you know what came for free.
4. Leave your water collector outside through a rainstorm. If you find when you return that the bottle is full, try putting a smaller funnel on.
5. Once you have an appropriate-sized funnel and bottle to get you through at least 24 hours of an average day, start recording rainfall levels whenever you empty the container.
6. Average rainfall calculations are usually reported by weather stations as daily or monthly volumes. Knowing the diameter of your collector, and by measuring the volume of water collected, you could estimate the amount of rainwater you would be likely to collect on a larger surface such as a roof or a larger water collector.

## Gray water

The water that runs down the drain from our shower is usually in much better shape than the water we flush down our toilet. The wastewater from the shower is called gray water, as opposed to black water, from the toilets. On its own, gray water requires little energy to treat, in order to return it to a pristine state. Once mixed with black water and other waste streams, those potential savings are lost. One solution is to use that water again, rather than send it for treatment. Because of the nature of both bath and kitchen wastewater, an ideal use for it is in the garden. Our shower water contains pieces of skin and other organic materials, while kitchen water is often filled with food scraps. Both of these are useful nutrients to gardens; however, there are certain risks involved with storing and using gray water that a person should be aware of before deciding to go down this route. It is often the case that the mixing of gray and black water happens almost immediately, so separating is something best thought of early in a new construction or large retrofit project, not as an afterthought.

The largest danger of using gray water occurs when it is left in a container for an extended period. As the water stagnates, the organic matter suspended in the water provides a good environment for harmful microorganisms, the source of disease. The best course of action is to use gray water immediately: as it is often used to water plants, simply have the gray water flow straight into the garden. It is easiest to think about this at the time when you are building or renovating, as the chore of separating and then applying the water is not typically worth the effort, except in areas where water is very scarce. It is advisable to take care which cleaning products are used if you decide to use gray water in a garden, as some chemicals can kill your plants. Look for products that are biodegradable.

## Composting toilets

If individuals decide to go down this route, and divert for immediate use all of their gray water, they could realize that the vast majority of their water is used to transport their feces to be processed. A person very concerned about water use would look at the porcelain fixture of most American households as a problem. There are several well-designed professional compost systems that have increasingly dealt with the problems of sanitation and hygiene that tend to accompany home-built versions. The reader is encouraged to explore the resources on Evilgeniusonline.com and look at professional models of composting toilets, rather than attempt their own. Low standards of cleanliness and hygiene plague home-built systems, and those done for "environmental" reasons risk tarnishing the name.

# Chapter 7

# Heating with the Sun

The sun is a gigantic nuclear reaction happening frighteningly far away (about 150,000,000 km or 93,000,000 miles away). If there is going to be a nuclear reaction going on, that seems like just about the right distance away from me for it to be happening. Nuclear reactions tend to produce large amounts of energy and astonishing temperatures, and the sun is no exception. Only a fantastically small amount of that energy ever gets anywhere near the Earth, for many reasons. For one, each of us is only ever on one side of the sun at a time and the rest of the time all that energy radiates the wrong way (for us anyway!).

Of the part of the Sun's energy that does get to our planet, about half gets through a thick layer of gases, called the atmosphere, before reaching the ground. Much of the energy that hits the ground is radiated back out into space, by the oceans for example. Even so, there is more than enough solar energy reaching us today in useful places to meet many of our energy needs, as long as we learn how to use it wisely.

## Our solar-powered society

The truth about the sun's energy is that it is everywhere, in almost everything around you, including the gasoline you're putting into your car and the wood in your house. What happened is that energy from the sun struck the earth and made it possible for things to grow, like plants and trees, and animals who ate the plants, and bigger animals that ate the smaller animals. This includes you and me! When this happened a long time ago and those plants and animals died, they fell onto the ground and were slowly buried by earth and water. Over time they were compressed, and those living things that had stored up the sun's energy a long time ago eventually turned into a gooey black substance (oil) that we now turn into petroleum products. When we dig it up and thin it out, it is sold as gasoline, diesel, and home-heating oil. Found alongside this oil is a gas—produced when you take living things (which used the sun's energy while they were alive) and put them away from oxygen for a period of time—called methane. It is commonly sold as "natural gas" when it comes from underground supplies formed long ago, but it can be produced from other sources too.

The problem with using up this stored solar energy is that the gases produced when we burn it are changing the atmosphere and the earth around us. It's also much better to make use of the solar energy that is striking us today than to rely on the stored energy from so long ago, because eventually it will run out. When that might happen is anyone's guess, but the theory is called peak oil (maximum rate of global petroleum extraction is reached); if you want to do more research there is plenty of information available. We'll just focus on the present day's sun for now, because there are literally thousands of different ways to put it to use. How you go about it really depends on what you are looking to do.

Different parts of the planet receive different amounts of sunlight at different times of the year, because the Earth is slightly tilted. The Earth spins on a slight (23.5°) angle off of what would be up from a horizontal plane formed by the earth's path around the sun. There's not really an "up" in space, so it's a bit tricky to get your head around, but the main thing to understand from the point of collecting solar energy is that because of this angle, the path of the sun through the sky will change through the seasons in predictable ways. (See Figure 7-1 for a visual representation.)

Living in some of the warmer parts of North America (or the world), you might often think that the most useful thing the sun could do would be to go away and leave some cool air behind. Solar shading is a building technique that creates shade in particular places in order to reduce unwanted heat from direct sunlight. It is particularly useful in locations where air leaks often occur, like windows and doors. By paying attention to how the sun moves, it is straightforward to reduce the cooling

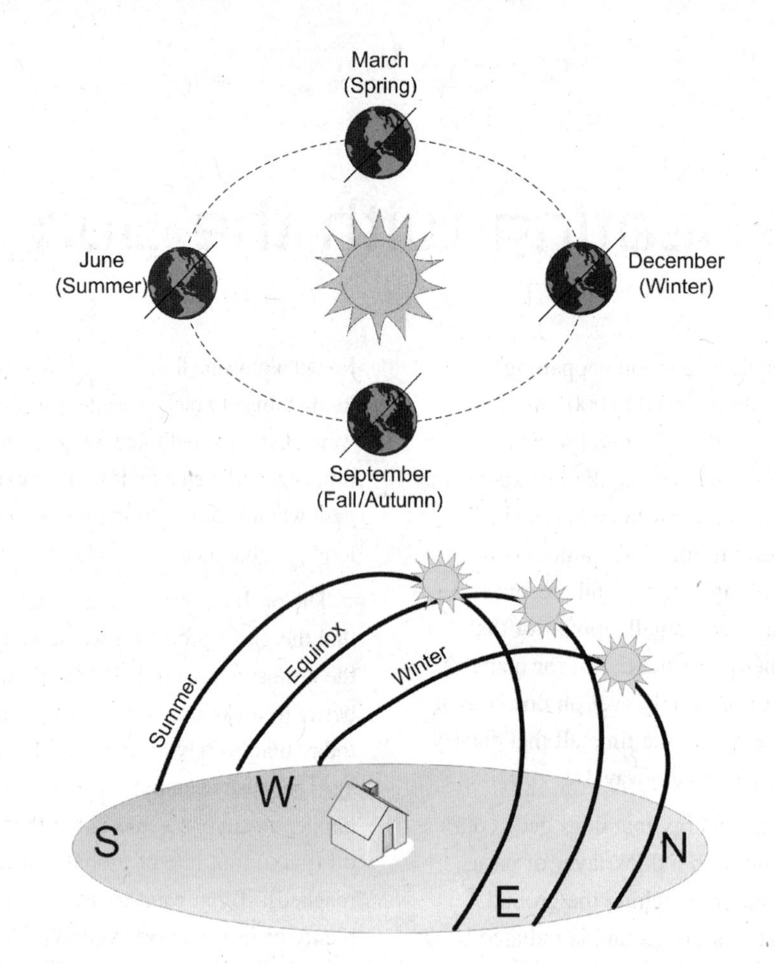


Figure 7-1 *How our slightly tilted earth results in the path of the sun through our sky changing through the seasons. Used with permission from Gavin Harper's* Solar Energy Projects for the Evil Genius. *McGraw-Hill, 2007.*



requirements of a space. Peak electricity demand in hot parts of the world tends to happen in the summer, as a result of air conditioner use, so besides shading, actually providing cool air seems like a common thing the sun could help us out with. Because of how heat machines operate (again more on that later), gathering heat is a useful step towards getting cool, so we concentrate on heat for now. Gavin DJ Harper showed us evil geniuses how to build a solar-powered cooling machine in *Solar Energy Projects for the Evil Genius* (check it out if you haven't already), and we have seen one that combines the sun's heat with water's ability to absorb and draw heat away when it evaporates.

In colder parts of the world, where the sun shines less often, the most important thing the sun is likely to be able to provide is heat. We saw in Chapter 1 that the average person spends about 14% of their energy budget on heating water. Too often that heat is provided by fossil fuel sources, when there are alternatives all around. Often, we just have to learn to use them better.

## Making good use of the sun

There are several things you can do to help maximize the amount of solar energy you can capture and put to good use. Not only is the sun frighteningly far away but it is also the thing that our planet is orbiting, which means it will move in the sky through the day, but in predictable ways. If you are in the northern hemisphere, the sun is primarily in the southern sky. So in the previous experiments, if you chose a good location it probably had some southern exposure, meaning that nothing was blocking the sky to the south. While we were only heating up a bottle of water, it is possible that you have already found a good location for other solar-based activities. Trees and buildings commonly block the sun out from particular locations; getting higher sometimes helps.

# Project 24: A Simple Solar Cooker

We are going to use some scrap materials found around your house to demonstrate how to capture a little of the sun's energy in order to heat some food or water. You can make this project as big or small as your materials permit (not everyone eats pizza so it's not a requirement to follow the pictures-exactly), and there are loads of freely available plans on the Internet for specific designs and improvements. For now, we're concerned with making sure the avid reader understands the concept of trapping the sun's warmth. We'll make use of this concept again later when we discuss environmentally sensitive building materials and energy-efficient building design.

**Figure 7-2** *Two old pizza boxes, about to be repurposed as the base of a solar cooker. Other than the black bottle (which we'll use in the next project), these are the main parts of a low-tech solar cooker: plastic wrap, tin foil, newspaper for insulation, and pizza boxes.*

## You will need

- One large cardboard or wood box. We will be cutting it diagonally, so a material that you can work with is best. Alternatively, build your own from some material that can form a right angle and triangular corners. (See Figure 7-2 for materials needed for a super simple version using a pizza box and two scrap pieces of cardboard as the walls.)
- One slightly smaller box of similar proportions to the larger one.
- Scissors and some tape.
- A glass or plastic window (in a pinch you could use plastic wrap, too).
- Shredded newspaper (or something similar) as insulation.
- Aluminum foil.
- A vessel to cook in (preferably black or painted black).

## Steps

1. Slice both boxes diagonally from corner to corner so that two corner pieces remain (note: save the second corner piece for a second cooker!), or assemble the triangular piece in place to hold the back of the pizza box upright. Think about how you will access the inside, to place and remove food or water, at this point. It could be through the side, or by lifting the glass off.
2. Place a layer of shredded newspaper inside one of the larger corner pieces (see Figure 7-3).
3. Line the inside of one of the smaller corner pieces with tin foil (see Figure 7-4).
4. Place the smaller box inside the larger.
5. Affix your glass to the outside box, making sure that a seal with the inner box is maintained (see Figure 7-5). Add more insulation if necessary.

**Figure 7-3** *Newspaper (insulation) inside a reused pizza box.*

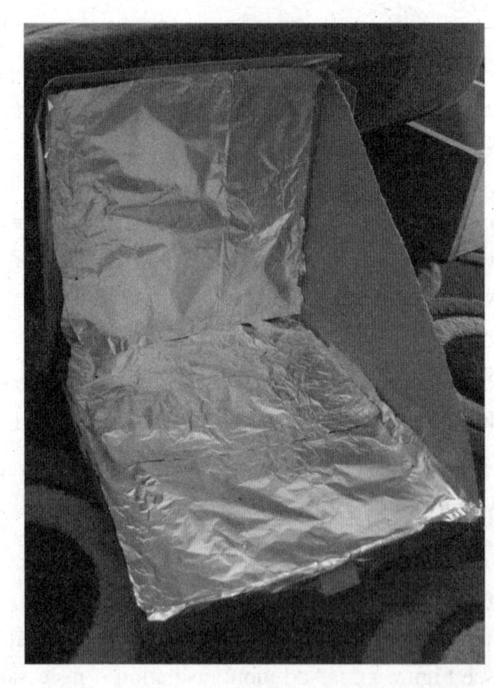

**Figure 7-4** *Tin foil lining the inside of the smaller pizza box, which will sit on top of the layer of paper insulation.*

**Figure 7-5** *The glass in this case is just plastic wrap, covering both boxes. A makeshift door (facing the camera) allows access to the interior to place and remove food.*

**Figure 7-6** *The completed pizza box solar cooker in the sun. The temperature inside reached a balmy 70°C (158°F) within 90 minutes on a partly cloudy day—temperature of 25°C (77°F)—in New Brunswick, Canada.*

6. Place your full pot into the cooker (with a thermometer); then close and seal the window to the box (see Figure 7-6).
7. Orient your solar cooker so that there is close to full sun on the clear angled wall of the cooker.
8. Cooking in the sun may take longer than a normal oven, so try to cook during the heat of the day.
9. If you were unsuccessful at reaching 65° C (150° F) with your water purifier, try placing the purifier in the oven and trying again.
10. Think about how to make your cooker more practical. Remember that both water and stones can maintain their heat for some time. Is it possible to cook at night over sun-heated stones? Don't be limited to cooking outdoors either; beautiful indoor kitchens could be built around solar cookers.

## Heating water

There are literally thousands of ways to heat water with the sun. Our earlier project showed how even the small act of painting part of a plastic bottle black can increase the amount of thermal energy captured from the sun. Our solar cooker then showed the importance of insulating those areas that would tend to lose heat, while allowing a section for solar thermal energy to be captured. We can apply both of these principles to the task of heating water. Water is a really useful thing to heat! Not only do we like to have it at a comfortable temperature to use when bathing and washing, but also we can use it to move heat around to more convenient locations and radiate it away into a room.

# Project 25: Simply Heating Water with the Sun

Our first water heating project combines the principles from earlier projects: basically, an insulated box containing a blackened vessel filled with water is used to capture the sun's heat. You could do this on a large or small scale; I am a big believer in doing things small before making them big, so we will consider both. Figure 7-7 shows a practically sized system using this principle. The main difference from the miniature ones that we built in earlier projects is that it has a piped connection to a showerhead below, to make some good use of the heated water. Shown in Figure 7-8 is an even simpler system that uses just a coil of black plastic tubing placed on top of an aluminum roof and again plumbed into an outdoor showerhead. Simple systems like this, which are useful in the summer when there is plenty of heat or in places where there is always a lot of sun, can be very cost-effective.

**Figure 7-7** *A batch solar hot water heater, with a metal tank inside an insulated box functioning much like a solar cooker, feeding an outdoor shower (the cracked glass is about to be replaced).*

**Figure 7-8** *A low-cost batch water heater that holds enough water in the coils of black plastic tubing on the roof to provide one or two outdoor showers at the end of a hot day. Photo taken at the Falls Brook Centre, Canada.*

**Figure 7-9** *The insulated solar cooker with a clear plastic bottle of water being purified inside. To the right are the water purifiers from an earlier project, which were all worse at capturing the sun's heat than this simple pizza box cooker.*

## You will need

- Either your solar cooker from Project 24, or a version of it, appropriate to contain the water holder you're planning to use.
- A vessel to contain the water you want to heat, preferably painted black. Because we're not trying to irradiate any of the water with the sun's rays, we don't need to worry about only painting half of the bottles black for this project, and can just spray-paint them all black.
- If you want to get fancy, or are trying to build a bigger system, you'll want to think about how to get the water out and into use. As in our examples, this could be by using a simple hose with a faucet or showerhead.

## Steps

1. The basic idea of the solar cooker is to use two insulated half boxes, with insulation between them, and a glass wall to point at the sun, as we learnt in the last project.
2. Mount the (preferably blackened) water tank inside the solar cooker (see Figure 7-9).
3. Point the glass end of your solar heater towards the sun for a few hours; then enjoy a purified glass of water (as we did) or a hot shower! Users of the larger system based on this principle (see Figure 7-8) have found that a warm afternoon in Canada is all it takes to have a nice shower at the end of the day.

# Project 26: A Circulating Water Heater

We didn't describe it as such, but the last solar water heater was what is called a "batch heater," because it heats one batch of water at a time. It is a pretty good way to heat a batch of fluid to a reasonable temperature, like 50–60°C (122–140°F), but you then have to worry about moving the batch water in and out once it is hot enough. The alternative is to heat moving water. Actually, this tends to be more efficient, because you can heat a larger quantity of water by constantly moving cold water into the area being heated by the sun. There are, again, multiplicities of ways to build a solar-powered circulating fluid heater, as well as an increasing number of commercial options. We'll start with a fairly simple home-built version in this project and then move onto some commercial systems later on. A professional system built on this principle is shown in Figure 7-10 on the roof of Sims House in Knowlesville, New Brunswick.

This project describes entries into a solar thermal design competition hosted by the Falls Brook Centre in New Brunswick. Each used a slightly different design, and the process of learning how to capture and raise the temperature inside the regulation 20-liter bucket most efficiently was the best part. None of these systems are terribly efficient, but they do the job reasonably well, were built from parts that included reclaimed materials, and encouraged learning. One (not the winner) of the systems has since been installed in an often-used shower and is performing admirably.

## You will need

- A few meters of black plastic pipe (or another color and some paint) was used for the system pictured, but other sizes will do. This can sometimes be scavenged or reused, particularly as the size isn't terribly important. The opposing team used copper pipe, which worked equally well but requires some extra tools and know-how.
- Several T- and L-shaped plastic connectors.
- A sheet of aluminum or other rigid backing of about $1 m^2$ (Figure 7-11).
- Some zip ties and (optionally) some silicon sealant.

**Figure 7-10** *Professionally installed solar hot water system in New Brunswick. The small electric panel to the right on the large thermal panel powers a pump which circulates fluid through the collector and into the tank inside. The flat-plate collector is similar to the ones built in the upcoming project.*

**Figure 7-11** *A reclaimed aluminum sheet, length of cut plastic pipe, and some hose connectors ready to assemble into a homemade solar water collector.*

- A drill with a bit sufficient to get through your backing material (or a hammer and nail as a hole-punch).
- A power supply for the pump. Our example uses the solar electric system from a future project and you can too: just jump ahead. This has the advantage of only powering the pump when the sun is shining, precisely when it is useful to circulate water through the panel. Otherwise, use a battery or a 120 V pump.
- A small pump with fittings for the size of pipe used and appropriate to the power source chosen.
- A bucket to use as a tank. For the competition, a 20-liter bucket was required and a limit of 1 $m^2$ governed the collector size.
- Some insulation and black paint to improve the performance of your system (optional).

## Tools

- Drill.
- Screwdriver.

## Steps

1. Start by laying out the hose path of your planned collector (see Figure 7-12) and thinking through the system. An important concept to remember when determining the layout is that hot water will tend to rise. This is called the thermosiphon effect, and your collector can use this to make the pump's job easier (some systems even replace the pump by locating their tank above the collector). To take advantage of this in your system, position the water inlet near the bottom of the collector, and the outlet closer to the top. Because we'll have a pump, it is not necessary to be more precise than this.
2. The idea is to have water in the hoses flow over as much of the aluminum sheet as possible, and be touching the aluminum if possible (see Figure 7-13). The aluminum acts as part of the collector if it is touching the hose, because it too will gather some of the heat and transfer it through to the water.

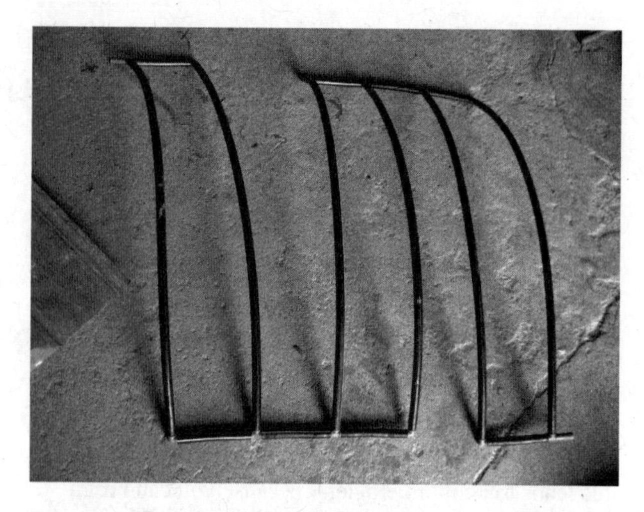

**Figure 7-12** *The pipe connected in its proposed layout before being attached to the aluminum backing.*

3. There is a lot of room for creative problem solving here. For example, we had planned to have our hoses travel up and down the front of the aluminum sheet six times, but found that we didn't have enough of the right sort of T and L connectors. Rather than go to the hardware store to buy more (a waste of fuel), we figured out that sending the water up and down only three time in a set of two hoses worked just as well.
4. Once you have finished planning, cut your hose to the right length and join it up in the right shape by connecting it together with the T and L connectors. You might need hose clamps, or pipe tape, to stop leaks once you test it out.

**Figure 7-13** *The proposed layout of the hose across the aluminum sheet. The inlet and outlet connections are positioned to take advantage of a mild thermosiphon effect, but there will also be a pump.*

Figure 7-14 *Zip ties fed through holes drilled in the aluminum hold the plastic pipe securely in place.*

5. Drill holes in the aluminum and use the zip ties to fasten the hose to the aluminum backing (see Figure 7-14). You could more firmly affix the tubing to the aluminum backing using some silicon sealant if desired.
6. You should now have a functioning collector (see Figure 7-15). Much of what we do from here improves its efficiency, but this is the "guts" of it.
7. Get longer lengths of hose that will feed into your tank and attach it to your pump and tank.
8. Fill the bucket with water and try to siphon the water through into the collector.
9. Angle your collector facing the sun and power up the pump (see Figure 7-16).
10. Depending on how bright the sun is, you could see results within a few minutes.

Figure 7-15 *The completed basic collector. The inlet is visible at the bottom right, the outlet at top left.*

Figure 7-16 *The homemade solar collector in action! It has been painted black and has some (reclaimed hot water tank) insulation behind it. The tank is kept (out of the sun) in the container doubling as a stand, and the pump is powered by the solar panel mounted on the fence behind.*

11. It is a good idea to test your collector out in the sun before painting it black or putting any insulation on it, just so you can have a real experience of how much of a difference the color black and some insulation makes to the overall efficiency of your system.
12. Spray or apply your black paint, and attach some insulation to the back of your collector to improve its efficiency.

## Alternative design

Another version of a solar thermal collector uses a wooden frame and some copper tubing to capture the sun's heat. The example in Figure 7-17 was built by students at Carleton North Secondary School and won the 2008 Energy Experience design competition at the Falls Brook Centre in New Brunswick. You'll notice from the image that it uses different principles to the last project. They chose to construct a wooden box and use copper piping, as well as a layer of glazing over top of, not unlike our earlier project

Figure 7-17 *Builders of the winning solar thermal collector check on its performance. The solar electric panel to the right powers the pumps, again hidden behind the thermal collector and doubling as a stand.*

Figure 7-18 *Mist is seen on the inside of the thermal collector, reducing its efficiency. Also notice how the copper tubes are neatly held in the center of the casing, so the heat they collect gets moved to the water instead of the case.*

of putting water inside a solar cooker. One of the problems they encountered, which you can see in Figure 7-18, was that some of the water evaporated from the pipes and formed a misty cloud on the inside of the collector. This reduced its performance, but they still got enough heat into the water to win. I'll leave it to you evil geniuses to plan another competition and come up with improvements to both these designs.

## Project 27: Solar Thermal Panels on Your Roof

Using solar thermal energy to supplement your current hot water heating source is an efficient and cost-effective method of getting familiar with larger-scale renewable energy systems. Home-scale solar thermal energy has the fastest cash payback of any of the home-sized renewable energy technologies covered in this book. The average 2008 model pays for itself within about 5 years, depending on your current fuel costs. As fuel costs rise, this number will get shorter and there are new programs by governments to subsidize the upfront purchase cost being created all the time. After 5 years, you will be collecting free energy to heat your water, and many systems have a warranty for up to 25 years.

Current models interact efficiently with a second heating source, such as mains electricity, by locating two heat sources inside the same water tank (see Figure 7-19). The electronic systems are smart and turn off circulation when the solar thermal collectors are colder than the tank, preventing your heat from rising up into space at night. The computer seamlessly turns a supplementary heater on and off to a schedule or based on a minimum tank temperature (the controller

**Figure 7-19** *An individual family-sized solar-electric hot water heater, installed. The electric element is midway up the tank, fed by electric cables from the right.*

is shown in Figure 7-20). Really, this is a painless way to reduce your energy costs at home, and have the experience of renewably heated water in your shower. If you are at all interested in having a green home, a solar thermal system is a very wise first project. It will function without maintenance for several years, and save you money. Real genius.

**Figure 7-20** *Temperature sensors in the tank feed data to this computer, which then turns circulation on only when the sun is shining, and turns on the electric heater only when needed.*

**Figure 7-21** *An affordable evacuated tube solar thermal collector mounted on the roof of a house in New Brunswick, and a solar electric panel on a frame on the ground in front of it. The solar thermal system is by far the more affordable option.*

The basic principles behind some commercial flat plate collectors are very similar to the systems we have already built. They have a fair level of efficiency and are often very cost-effective. The students from Carleton North who won the design competition, made a system that in many ways resembles the increasingly affordable evacuated tube collectors (see Figure 7-21). Evacuated tubes (detail in Figure 7-22) use the same principle of a dark area (or tube) surrounded by insulation in order to trap the maximum amount of thermal energy, but they use a small amount of a highly specialized fluid instead of water. Specialized heat transfer fluid runs inside thin tubes that are mounted inside a glass tube that has had all the air sucked out—a vacuum, which is a fantastic insulator. At one end of the evacuated tube (the end of the tube on the left in Figure 7-22, usually mounted at the top when installed) the inner heated fluid emerges from the vacuum and is transferred to another fluid (often glycol). The working fluid circulates through the heat exchanger in the tank, when the temperature probe indicates that the collectors are warmer than the tank.

Figure 7-22 *Evacuated tubes before installation. The end of the tubes on the left in the picture is where the heated fluid emerges from the vacuum.*

Figure 7-23 *Parts of an evacuated thermal collector frame mounted on a roof of a house, not quite ready for the tubes to be installed.*

Installation on your house is likely to require professional assistance, though there may be some areas where you might assist. Professional assistance will be crucial to ensuring the frame (Figure 7-23) that holds the tubes is secure enough to withstand weather and wind gusts, is angled correctly for your location, and the system is connected to your existing water line correctly.

# Project 28: Solar Thermal Hot Water on the Roof of Your House

Depending on where you are and what sort of roof you have, the process of mounting a professionally built system will vary. Hire a professional if going with a ready-made collector; both it and your house are valuable. If you are installing on a roof that isn't as valuable, think about whether you need a ready-made system. The system from our last project was easily mounted (Figure 7-24) on a rooftop as a supplementary heater for a shower.

Figure 7-24 *The homebuilt solar thermal panel from the previous project. It has been simply mounted in a more permanent place on a roof as a supplementary heater for a shower.*

## Frame and tubes

Mounting frames of this evacuated-tube system consist of light aluminum bars that bolt together and are secured to the roof. Supports run across the top and bottom to hold the tubes. Special holders for individual

**Figure 7-25** *Specialized plastic tube holders (held right in foreground) are attached to an aluminum mounting-bracket (leaning on picnic table).*

evacuated tubes are attached onto an aluminum support member (Figure 7-25) in a test run on the ground to make sure it will all fit. A lower support member is bolted to the frame on the roof (Figure 7-26), before the row of specialized tube holders pictured (Figure 7-27) are individually put in place, with each evacuated tube. These will be different on each and every system and not interchangeable.

**Figure 7-26** *Supporting bracket for the evacuated tubes is attached to a roof-mounted frame.*

**Figure 7-27** *Lower support member of the evacuated tubes, ready to be lifted onto the roof.*

If you are not scared of heights, you might want to help installing the evacuated tubes into the finished frame. The tubes will probably come packaged in a box—carefully lift one out (Figure 7-28). Holding the evacuated tube away from your body and anything sharp, because it shatters easily, climb onto the roof (Figure 7-29). Orient the tube so the end of the tube where the superheated fluid emerges is pointing up, where it will be mounted into the heat exchanger across the top of the frame (Figure 7-30). Some systems are

**Figure 7-28** *Gently lift the evacuated tube from the box.*

**Figure 7-29** *Carry the evacuated tube onto the roof, holding it away from sharp objects that could shatter the glass.*

**Figure 7-31** *Apply a small amount of thermal paste to improve heat transfer.*

designed for a little paste to be smeared on the tube before mounting to ensure heat transfer (Figure 7-31): put some on if so. Place the top of the evacuated tube into the heat exchanger at the top of the frame (Figure 7-32). Push the tube securely into its socket, holding it with both hands and pushing straight in (Figure 7-33). With the top end securely in place, retrieve the lower mounting piece from a (hopefully) convenient location (Figure 7-34). Attach the plastic mounting cap to the lower end of the evacuated tube (Figure 7-35). Secure the capped tube to the frame with bolts and tighten (Figure 7-36). That's it. Go back down and get the next tube (Figure 7-37). You'll only have to repeat the process 20 times or so for each collector!

**Figure 7-32** *Slide the evacuated tube into the mounted heat exchanger.*

**Figure 7-30** *Correctly orient the tube so the thin end (where the fluid emerges) will fit up into the heat exchanger.*

**Figure 7-33** *Grasp the tube firmly and slide straight up into place.*

**Figure 7-34** *Retrieve the special plastic mounting bracket from a convenient location.*

**Figure 7-35** *Attach the mount to the bottom of the evacuated tube.*

**Figure 7-36** *Insert mounting bolts and tighten.*

**Figure 7-37** *That's it! Go back down and get the next tube.*

## Tanks and pipes

You're almost certainly going to need a professional plumber, or at least someone who has done plumbing work before, to do this part for you. Leaks are common when new plumbing is installed, so keep a look-out the first few days after your new system is installed. Your plumber may not take the time to insulate the pipes carrying the hot water from your collector to the tank; you'll want to make the effort. There are tube-shaped packages of insulation you will find at the hardware store or through your installer. You'll almost certainly be getting a new tank, one that is superinsulated and often with a secondary heat source. The thermal collector we just saw getting installed will be plumbed into the tank pictured in Figure 7-38. The tank can operate completely on an auxiliary (electric in this case) heat source while your thermal panel is being installed, so there is no need to worry about going without hot water for long time while it is being installed. The solar-heated water will be fed in through the connection visible at the bottom of the tank in Figure 7-38; it was shown with plumbing in place in Figure 7-19. The computer to the right of the tank (also shown in Figure 7-20) controls when water is circulated through the tank and when the auxiliary heater turns on.

Figure 7-38 *Solar hot water tank, before plumbing from the thermal collector is installed.*

You'll be in charge of programming the computer's settings, so stop and think about how you use hot water. Is it really necessary to have 80° C (175° F)water all the time? Try lowering the temperature; then experiment with the timer. Remember that the sun will shine brightly at predictable times of the day, so setting the auxiliary element to come on in the morning before the sun rises (if you don't shower till later) might mean that you have electrically heated water in your tank already, when the sun comes out. During the day a lot of good solar heat would go to waste because it would have nowhere to go—the tank would be hot already. There's nothing wrong with using electricity, as we'll see in the next chapter, but if you didn't shower before the sun came out, all that electricity was not wisely used. No need to get up earlier—the opposite in fact. With your new solar thermal heater, you now have an ecological reason (if you wanted one) to stay in bed a little longer: you are waiting for the sun to heat up the water before you wander over for a real feel-good shower.

# Chapter 8

# Water and Air

We are surrounded by both water and air and the majority of the surface of the Earth is a boundary between the two substances. Each on its own has remarkable and useful qualities, many of which we use to a great degree already but which we could also stand to use more of. We have examined air as an insulator and a conductor of heat, and as a gas that expands when heated. We have also examined water as a heat store through time—a sort of battery—and something that can provide cooling. In this chapter we see how water and air can be used in a number of additional ways to provide useful work, heat, and electricity.

## Water power

Hydroelectricity provides a huge percentage of global energy, often through large systems built on major rivers. There are several devices commonly employed to turn moving water into a force suitable for performing useful tasks. The oldest is probably the water wheel, which has paddles attached to the outside of a large wheel. In pre-electrical times the moving wheel was enough to be of use for grinding corn or milling wheat for farmers.

There are at least two configurations of the water wheel: the water goes over the wheel or under the wheel. Having the water go over the wheel (called an overshot wheel) might involve changing the river a little more than going under the wheel (called an undershot wheel), but the trade-off is that it is possible to get the wheel turning with more force using an overshot wheel. Whatever changes do need to be made to the river for an overshot wheel will not likely be significant. Some undershot wheels, like one of the kits we explore, can even just sit right on top of the river.

Some version of the same idea that lies behind a basic water wheel lies behind all sorts of systems built today: some pumps have a prime mover that resembles a water wheel and can be used in reverse to generate power, and some custom-built systems have slightly modified paddles and use nozzles to direct the water at all parts of a wheel simultaneously, to maximize the amount of power extracted from the falling water.

## Project 29: A Model Water Wheel

Water wheels have been used throughout the world for a long time and there are countless examples of out-of-use hydro sites in small towns and villages in the United States, UK, and around the world (Figure 8-1). Often this is because towns and villages have traditionally gone up around rivers, both because they are good sites for river wheels and because rivers provide transportation and other amenities. It might seem quaint to look at this now, but we should remember that the alternative to big projects (which can have large environmental impacts) could be many smaller projects that have the potential to

Figure 8-1 *An old dam outside Florenceville, New Brunswick, which once powered a sawmill using a large water wheel. The wheel was fed through the chute, barely visible in the top left corner of the photograph.*

generate electricity and power for other purposes on a small scale for local consumption. Often these projects need not be complicated, but even when they are more complicated, modern systems are often based on the same principles as the model we build here.

An example of a slightly advanced version of the simple water wheel concept is the hydrogenerator constructed at the Centre for Alternative Technology in Wales, UK. This grid-tied hydrogenerator station has the turbine turned so the axis is vertical, and four nozzles of water are aimed to hit all four (custom-engineered) paddles simultaneously from four directions instead of from one. It is the same idea as the water wheel we will explore, just adjusted to ensure a higher efficiency.

## You will need

- Four small rectangles cut from firm cardboard, each side a little shorter than the length of a cork from an average wine bottle.
- A cork from an average wine bottle or equivalent.
- Two sewing pins or screws.

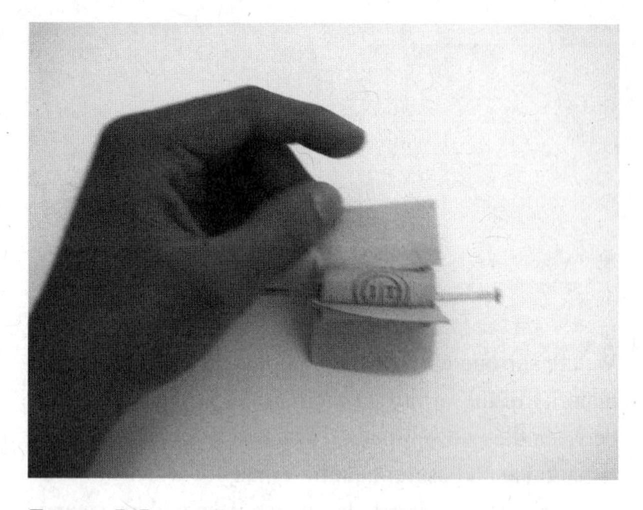

Figure 8-2 *Pushing the last paddle into a slit in the cork. The axles, made of screws, are already attached.*

## Steps

1. Cut four slits equidistant around the outside, running lengthwise down the cork.
2. Round two of the corners on one side of each of the four pieces of cardboard.
3. Slide the rounded ends of the four pieces into the four slots in the cork (Figure 8-2).
4. Put the two pins in either end of the cork.
5. Rest the pins on your fingers and have the water strike the paddles (Figure 8-3).

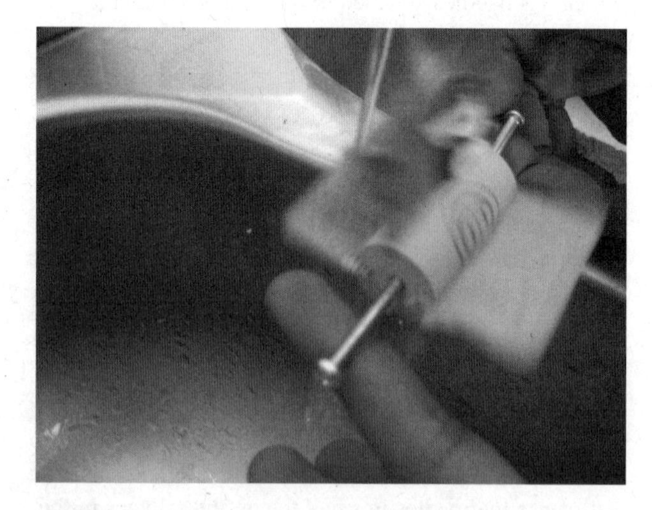

Figure 8-3 *Testing the water wheel, held on a thumb and finger, using a kitchen tap.*

## Variations

Mounted in even a simple frame it is possible to get real work from this simple machine. Attached to a 1.5 V motor it can recharge a small battery to power a light and, if it is fed from an elevated rainwater barrel draining into your garden, you can catch and make use of some spare gravity-fed water energy!

# River courses

There are generally two kinds of river-based hydro systems, both of which modify the course of natural water flows to harness some of the naturally flowing energy. A dam system blocks the flow of a river completely and sends the entire contents of the river through a turbine, which turns an electric generator. A run of the river system sends only part of a river through the turbine and leaves the rest of the river undisturbed.

Hydro systems can be good for providing a stable source of non-polluting electricity in large quantities, but sometimes their impacts on the landscape, including people and animals, owing to the creation of large reservoirs to hold the water back (Figure 8-4) and during their construction, can be quite enormous. The well-known Three Gorges Dam in China is just one example of a dam-based project that displaced millions of people to create a holding pond for the enormous volume of water that would be fed through the hydroelectric generators. Similar examples can be found constructed throughout North America and Europe, often further into the past so their impacts are further from recent memory.

A third type of system for providing energy from water uses water and gravity together as a type of battery and is among the most efficient types of batteries known. A water battery uses two separate reservoirs that are at different elevations; e.g., one higher up a mountain than the other. The top reservoir can be pumped full, using the water from the lower reservoir—e.g., by using electricity from the grid during off-peak hours or from a wind system. The top reservoir can then be emptied through a turbine to generate electricity at times when it is needed, such as during peak hours or when the wind has died down.

**Figure 8-4** *Even the small holding pond (pictured) built up behind the dam shown in Figure 8-1 dramatically changed the existing landscape and river.*

Water batteries, while efficient, are only really suitable for a small range of places, where reservoirs or potential reservoirs already exist naturally. Disturbing those places could have a potentially large impact. A water battery also requires another source of power to move the water against gravity to the higher reservoir. In some cases surplus off-peak electricity can be used and then fed back onto the grid when needed. There are also other, non-electric, devices that can move water up great heights using only the force contained in moving water, without even blocking the flow of the rest of the water. The ram pump uses the pressure built up in a column of water by a stream of moving water, to push some of the water up great heights. The Online Resources have links to a ram pump project you could try, and an upcoming project explores a similar way to do the same job, which is easier to demonstrate and quite neat too—the spiral pump.

# Project 30: A Model-Sized Water Wheel

To get a better sense of how water in a moving river can be used without modifying the river in any way, it is educational to construct a more practical model-sized water wheel. The kit pictured, made by Walter Kraul Gmbh (www.spielzeug-kraul.de/), is sold through many retailers around the world, and is an example of the fine sorts of learning tools that can be crafted from long-lasting wooden pieces.

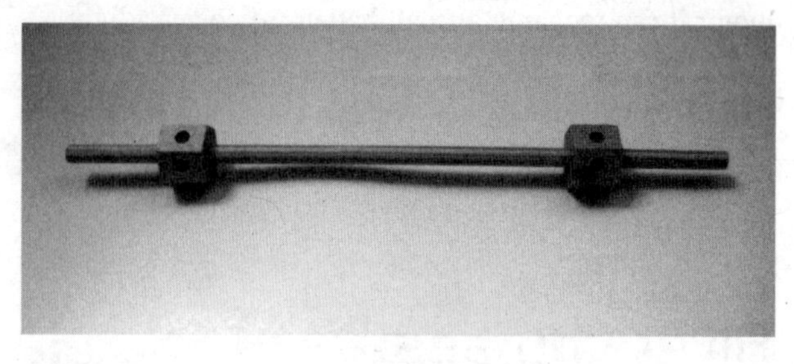

Figure 8-5 *The wooden axle and pre-drilled hexagonal hubs from the Walter Kraul kit.*

## You will need

- To order the Fluß-Wasserrad kir (River wheel at the Centre for Alternative Technology's store www.cat.org.uk), or a similar wooden wheel, or for the real genius, to construct something similar from spare parts. The Online Resources have some links to larger water wheel projects made from PVC pieces.
- The kit requires no extra parts and comes with complete instructions that are easy to understand. The company sells a variety of interesting wooden projects to experiment with and put your waterpower to use, including a model grain grinder and other useful tools. As their site is in German, it may be easier to search down an English-speaking retailer nearer you.

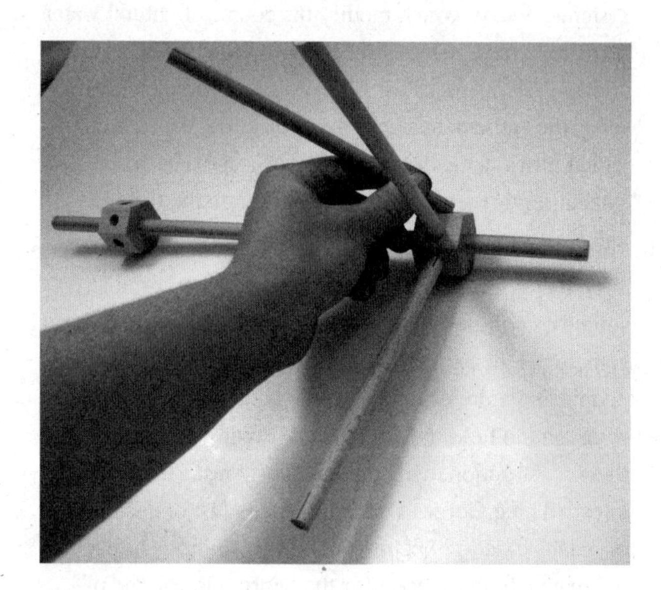

Figure 8-6 *Inserting the outriggers into the hub.*

## Steps—assembling the water wheel

1. Using the wooden kit as an example, slide two notched and drilled hexagonal hubs onto the axle (Figure 8-5).
2. Insert the 12 thin dowels (outriggers) that will hold the paddles (Figure 8-6).
3. Attach the paddles to the dowels using the provided elastics (Figure 8-7).

Figure 8-7 *The first of 12 thin boards to be attached to the outriggers with the supplied elastics.*

**Figure 8-8** *String wrapped around the outside of the outriggers, and bead-shaped stabilizers in place.*

4. Once you have attached two pieces of wood to each pair of dowels, wrap a string around the outside of the paddles for support, and place the bead-shaped supports in place, to keep the paddles from sliding down the axle (Figure 8-8).
5. Use the supplied wing nuts and bolts (Figure 8-9) to assemble the base.

**Figure 8-9** *The wing nuts are a tight fit on the base.*

**Figure 8-10** *The completed river-wheel kit, ready to be tested.*

6. Finally fix the wheel into the base and attach the drive pulleys (Figure 8-10).
7. Test your water wheel out in a slow flowing river, making sure to put heavy rocks on the base to keep it from floating away (Figure 8-11). We will look at useful things to do with your moving wheel in a later project and in other chapters.

**Figure 8-11** *Don't forget the rocks; the water wheel floats a little until the wood gets wet, and can get carried away easily. Because the paddles are widely spaced, the motion of the wheel is not completely smooth, and observers get a very good sense of when each paddle develops its force from contact with the water.*

## Online Resources

- www.clemson.edu/irrig/equip/ram.htm— W Bryan Smith's ram pump project site.
- www.green-trust.org/2000/rampump.htm—a long-running weblog and resource for new and old technologies.
- builditsolar.com/Projects/WaterPumping/waterpumping.htm—in addition to providing information about many types of different types of water pumping systems, Builditsolar is a fantastic resource for the do-it-yourself renewable energy enthusiast.

**Figure 8-12** *A commercial electric air compressor and high-pressure tank, used on construction sites as a portable power source, like a chemical–electrical battery.*

## Under pressure

An air compressor can be used to create many types of useful work and is a common industrial tool. Most of us are at least familiar with medium pressure air—the tires in your car. Air can be used in other ways with a fair degree of efficiency: e.g., by being put under pressure using a compressor, and then used as it is uncompressed.

Pressurized air is commonly used as a power source and effective battery in the construction industry (Figure 8-12) to power tools which have fewer moving parts than their electrical counterparts. With proper safety features in place, compressed air also has the potential to be a low-impact and high-efficiency storage medium. There are large-scale projects by utilities in Germany and the United States to use compressed air to store electricity from the grid at periods of low demand—like the night—and use the stored compressed air to improve the efficiency of gas-powered turbines by two-thirds, and in some proposed projects to feed electricity directly back on to the grid. The economics and practicality of large-scale pressurized air storage projects depends on the correct geological formations to store the compressed air: for example, underground salt caverns are often useful as they provide large spaces that can be sealed to contain pressurized air.

It is not hard to explore the concept of making pressurized air useful, even on a small scale: just blow up a children's balloon or pump up the tire of your car and then let the air out into the path of a small wind turbine (which we'll soon build).

## Hot air

We will explore another way in which air can be used when discussing the Stirling engine (see chapter 9), and during the solar projects. The Stirling engine depends on air expanding when heated to run; the same "feature" of air often leads to losses in air compressors, where putting air under pressure leads to heat losses. Conversely, when the air is expanded, it cools down and draws heat in. The efficiency of compressors in factories that power large systems and provide compressed air for cleaning and other functions can be improved dramatically by providing bigger heat sinks (such as using it to heat indoor spaces) and a cooler air intake temperature (such as in taking the air in shaded areas).

Some have started to think about improving the efficiencies by storing the heat produced by compression separately from the air, and reusing it to preheat the air as it is expanding and doing work, thus preventing energy losses. This is being done on a large scale, as well as potentially on a smaller scale too and, in combination with a flywheel, is being packaged by one company (see the Online Resources) as a non-chemical battery-based uninterruptible power supply (UPS).

There is a lot of potential for innovation in this area, especially now that we have very strong and light materials, such as carbon fiber, to build tanks. The underwater diving community has a history of portable and relatively lightweight tanks, which some have already put to use powering vehicles (see the Quasiturbine project and homebuilt car at Evilgeniusonline.com). Though there have been promises of air-powered cars since the 1990s, and new projects seem to get announced all the time, they haven't made it onto the market fast enough.

It should be noted that high-pressure air can be extremely dangerous to work around, so it is very important to have professional or at least knowledgeable advice when working with these sorts of systems.

## Online Resources

- www.activepower.com—a commercial non-chemical, battery-based uninterruptible power supply that uses stored heat to improve the air pressure conversion efficiency.
- quasiturbine.promci.qc.ca—an innovative engine that runs off of compressed air without using traditional piston or turbines, rather a combination of the two, and which has been used to power a small go-cart.

# Project 31: The Spiral Water Pump

The spiral water pump is quite an old idea that hasn't had much widespread use. It demonstrates several interesting concepts: water piston, air pressure, and air and water transmission over distances. There are two parts to this project: the first is a water wheel, the one from the last project will do nicely to demonstrate; the second part is to build a spiral pump and affix it to the water wheel. When completed, we will send the compressed air and water up a hill, and separate the compressed air and water, so they can each do some useful work. The compressed air can be used to turn an air motor and the water can be used either as a source of running water or sent down the incline through a turbine.

It is useful to understand the concept behind this pump before starting to build it. The pump intake and compressor are both formed by a large spiral that is attached to the side of the water wheel. The spiral can be made out of rigid plastic pipe. The end of the spiral at the large end of the spiral should be positioned so that it gathers water as the moving stream turns the wheel. As the wheel turns with the spiral attached to it, the spiral gathers water during the first half of the rotation, then air during the second part of the rotation. Each rotation of the wheel pushes successive alternating batches of air and water into the spiral, which are squeezed into a tighter and tighter space in the parts of the inner spiral that have progressively smaller diameters. Because water doesn't compress very easily, the air compresses in the spiral.

At the center of the spiral, the alternating air and water "pistons" inside the plastic tube are turned 90° from the wheel and sent through a pipe up an incline. Air expands to push pistons of water up an incline, where the water and air can be separated. There may be enough air pressure to turn an air compressor and generate electricity. The water can either be used for other purposes, having been pumped up an incline, or can be sent back down the incline and used to generate power.

## Online Resources

- lurkertech.com/water/pump/tailer/—Chris Pirazzi has collected a wealth of interesting and hard-to-find material on water-powered water pumps, including the spiral pump created by Peter Tailer and a mirror of Peter Morgan's hard-to-access tripod site on large-scale spiral pumps.
- www.wildwaterpower.com/—Denis Buller's site on his inexpensive homemade spiral pump.

## You will need

- A medium-sized water wheel with a base. The pictured example (Figure 8-13) uses the water wheel from the previous project.
- Some flexible plastic tubing.
- A frame for the spiral pump that will enable it to stay on the side of the water wheel: e.g., long wooden boards and rope.
- Some right-angle hose connectors appropriately sized to the plastic tubing.
- A longer length of tube, in which the compressed air will expand and push the water up an incline.
- Optionally, a scoop to increase the water caught by each revolution of the spiral pump.

## Steps—the spiral pump

1. The pump is a thin strip of the plastic tube held in a spiral. Coil the plastic and attach supports flat against the tubes with rope or anything else handy; bolts and wooden supports can be used on a larger scale. The Online Resources has links to pictures of larger versions.
2. It is possible to use a fairly lightweight method to attach the spiral to the wheel axle (Figure 8-13), but a permanent installation should consider durability. The tube on the large end of the spiral should point downstream and for maximum efficiency could include a scoop.
3. As the wheel is turned by the moving water, the open end of the spiral tube will be pushed underwater (Figure 8-14) and gather an amount of water. Once the open end of the spiral rises to the surface (Figure 8-15), the column of water in the tube has a column of air up against the water.
4. As the diameter of the spiral shrinks, the alternating columns of water and air have less space, and the air compresses (Figure 8-16). From the center of the

**Figure 8-13** *Demonstrating how a spiral pump works using the water wheel from an earlier project. When finished, the tube at the center of the spiral will turn in a sliding-coupling at a right angle and send compressed air and water up an incline, without disturbing the course of the river.*

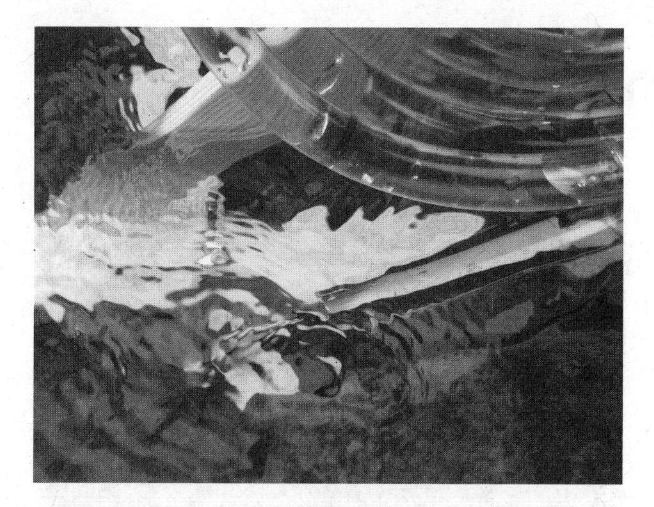

Figure 8-14 *The open end of the tube is pushed underwater by the wheel turning.*

spiral, attach a loose-fitting right-angle pipe fitting and a long tube to deliver the compressed air and water up an incline, where they can each be recovered and used.

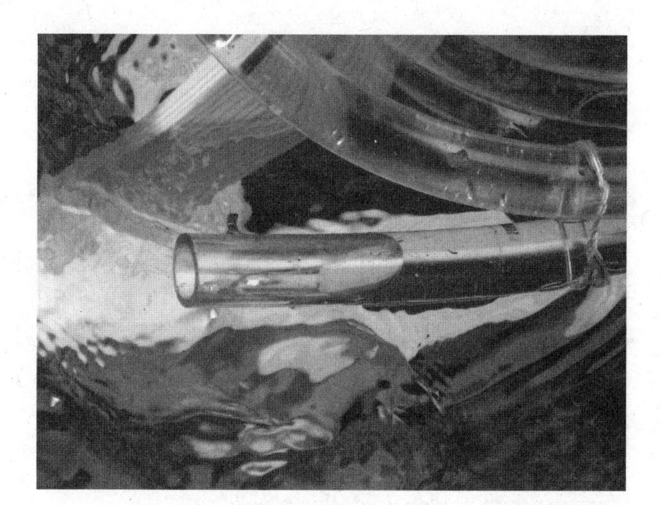

Figure 8-15 *As the wheel continues to turn, the open end of the tube is lifted out of the water, and starts to gather a column of air.*

Figure 8-16 *Alternating columns of air and water gathered in the spiral pump. As the diameter of the spiral gets smaller, the air is compressed. The alternating columns of water and compressed air can then be turned away from the spiral and sent up an elevation to do work.*

## Heat and air

A good place to start exploring compressed air is by looking at the primary source of inefficiencies. As much as compressed air can be a good way to store and use energy in ways that can be put together using relatively benign materials, inefficiency of the overall system could work against it working effectively. Compressing air creates an enormous amount of heat and that is a major source of perceived loss in pressurized air systems. In another application, the heat generated is also incredibly useful, like in a diesel engine, where pressure is used to ignite fuel inside the cylinders of the engine.

# Project 32: Making a Fire Piston

The fire piston is a very old idea, recorded by Europeans passing through South-East Asia around 1500. The device uses a small cylinder and airtight piston to ignite a small piece of timber, reportedly used to light cigarettes when "discovered" by the Europeans. Several budding entrepreneurs are selling ready-made versions of these devices as educational and useful items and there are videos of them in operation on the web—see the Online Resources for some links.

The pressures generated by a fire piston can be very dangerous and are beyond the design limits of the common materials that can be used to build a demonstration version. The devices can also be made from more natural materials, but they tend to be somewhat more difficult to build. A safer experience will almost certainly be had by purchasing one of the beautiful ready-made models (see Figures 8-17 and 8-18), which demonstrate, without much danger, the concept of using pressurized air to provide sufficient heat to create a burning ember. The reader is advised to read the complete instructions on the web and take appropriate safety steps before proceeding with this project.

**Figure 8-17** *A small Plexiglas fire piston, available from firepiston.com. Image courtesy of firepiston.com.*

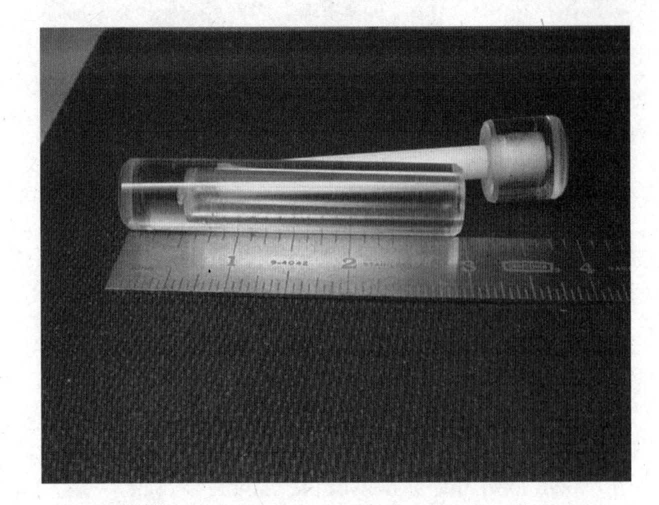

**Figure 8-18** *A larger Plexiglas fire piston, available from firepiston.com. Image courtesy of firepiston.com.*

## Online Resources

- www.wonderhowto.com/how-to/video/how-to-make-your-own-fire-piston-222301/
- wildwoodsurvival.com/survival/fire/firepiston/rbmodelt/—Rob Bicevskis' excellent site.
- www.firepiston.com/—a safer alternative to building your own fire piston.

Many thanks to Rob Bicevskis for providing images and guidance for this project.

## You will need

- Patience, time, and an Internet connection to research options and improvements to the system outlined.
- A cylinder. About 9 inches (22 cm) of half inch (12 mm) CPVC tubes and the same-sized end cap to form the cylinder is suggested in the model T project. An alternative is to use a metal tube with

a smooth inside finish and outside bolt threaded end and screw on cover, also available from most hardware stores, or a polycarbonate cylinder.

- A piston. For the model T version, a wooden dowel that fits tightly inside the cylinder but is longer than the cylinder (11 inch or about 30 cm is suggested to enable a comfortable handle). As an alternative, a smaller metal version could use a 4 inch long metal bolt.
- An O ring that fits tightly into a groove you will make on the piston (either dowel or bolt), which forms a seal inside the cylinder.
- A space at the end of the piston to hold a small amount of starter material that will be set alight by the heat generated by compressing air. For the PVC model, a smaller PVC end cap that will be modified to hold the flammable starter material. A small indent should be drilled into the bottom of the bolt for the metal version.
- Some flammable starter material (such as cotton wool or small pieces of timber).
- Possibly some grease for the piston.
- A PVC version could use various CPVC T junctions to form a handle for the model T type design and that will make for a safer experience overall, as well as dabs of CPVC cement and epoxy glue. See Rob Bicevskis' site for a complete list.

## Steps

1. There are three important parts to the fire piston: the cylinder, piston or plunger, and a strong and secure O ring to ensure a pressure build-up.
2. Creating a cylinder from CPVC tube is a relatively trivial matter of firmly affixing an end cap to one end of a length of tube with epoxy glue, which will be available with the CPVC tube at your local hardware store (see Figure 8-19). An alternative method is to use a metal tube from the nuts-and-bolts aisle with a screw end and securely screwed on metal cap.
3. The PVC model includes a T section that can be attached to the top of the wooden plunger (see Figure 8-20) to create a convenient storage space for Vaseline and starter material.

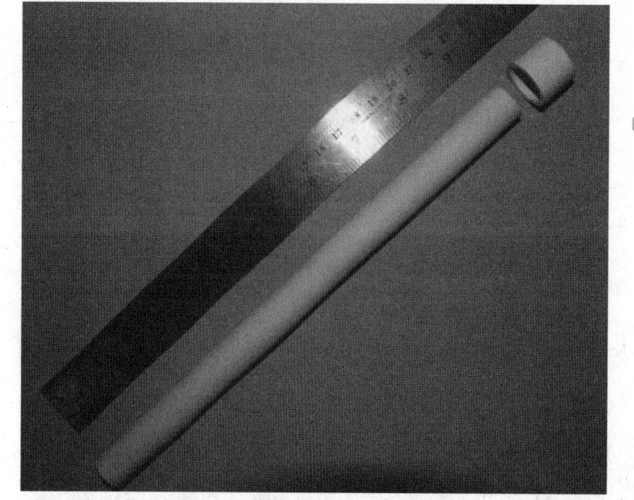

**Figure 8-19** *Length of PVC pipe and cap used for the model T fire piston. Photo courtesy of Rob Bicevskis.*

4. If using a wooden dowel, it may have to be thinned down a bit to fit inside the cylinder. It should not be airtight as the O ring will do that. Try to avoid the need to make a bolt smaller than it is.

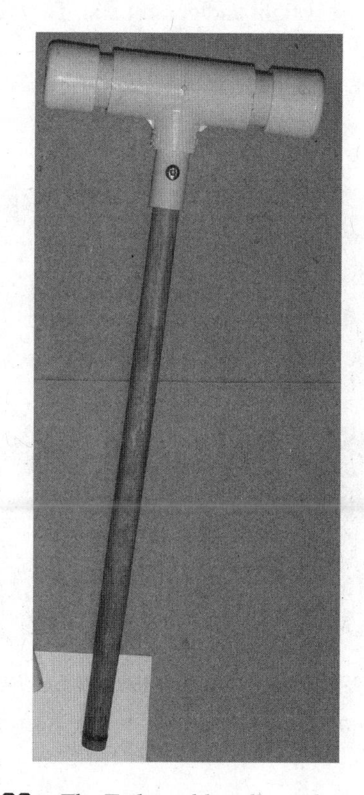

**Figure 8-20** *The T-shaped handle with storage compartments in the short ends of the T. Photo courtesy of Rob Bicevskis.*

**Figure 8-21** *Making the groove in the wooden dowel using a piece of string and some sandpaper. Photo courtesy of Rob Bicevskis.*

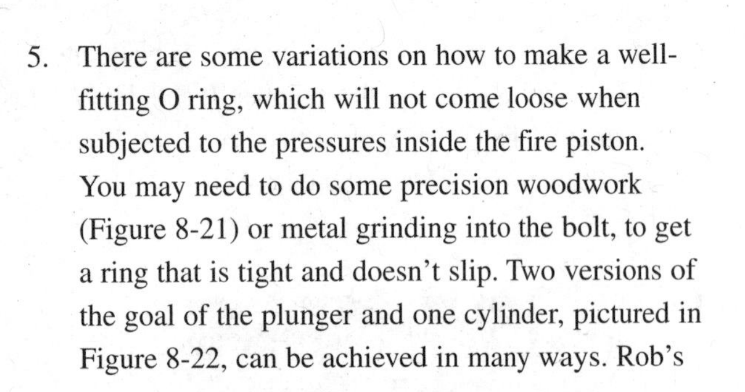

5. There are some variations on how to make a well-fitting O ring, which will not come loose when subjected to the pressures inside the fire piston. You may need to do some precision woodwork (Figure 8-21) or metal grinding into the bolt, to get a ring that is tight and doesn't slip. Two versions of the goal of the plunger and one cylinder, pictured in Figure 8-22, can be achieved in many ways. Rob's

**Figure 8-22** *Two pistons on the left and PVC cylinder on the right. Photo courtesy of Rob Bicevskis.*

**Figure 8-23** *Detail of the bottom of the piston, showing where the small amount of fire-starter will sit. Photo courtesy of Rob Bicevskis.*

site is a good start for more complete directions and links to other similar projects.

6. A hole drilled in the bottom of the piston (Figure 8-23) holds the timber that will be ignited by the pressure of compressed air.

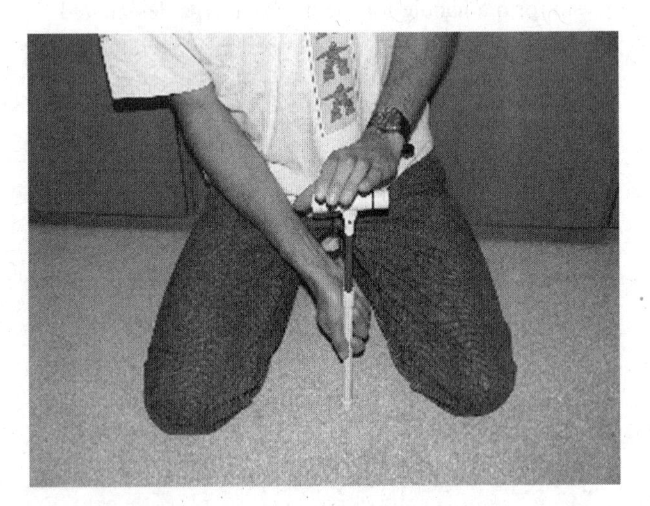

**Figure 8-24** *The completed model T fire piston. Push down on the T while it is the cylinder; the O ring forms an airtight seal that results in air compressing at the bottom of the piston, setting the fire-starting material alight.*

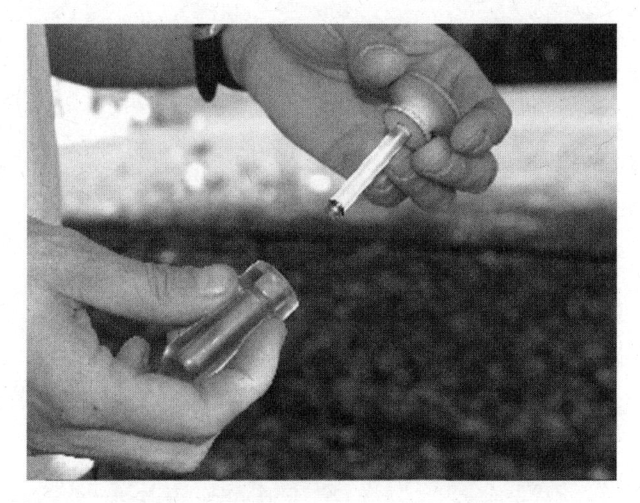

**Figure 8-25** *A smaller fire piston with its ember burning. Photo courtesy of Rob Bicevskis.*

7. One version of the fire piston should look like Figure 8-24 when complete. Fill the holder with a small amount of timber, grease the plunger, and in a swift motion push the plunger into the cylinder. Make sure you are ready for the reaction of the force of compressed air igniting the timber, which will push back at you. You can then remove the plunger and use the burning ember to light a fire! Another beautiful fire piston, also made by Rob, is shown in Figure 8-25, with its ember glowing.

# Chapter 9

# Heat and Power

There is the potential to use heat for a wide variety of purposes. The most common ways to use heat are space and water heating; however, heat can also be stored, as was seen in Chapter 7, or used to create useful work.

Although normally we think of heat in the high-temperature form found in a gasoline engine or a steam turbine, even relatively low-temperature heat like that found in a solar thermal panel can be useful. Theoretically, work can be extracted from any temperature difference; in practice, obtaining useful amounts of work from relatively small temperature differences is challenging. There are many different ways in which this is practically done. A device that turns heat into another type of useful work or vice versa is called a heat engine. Steam engines, air conditioners, and heat pumps are all heat engines, as is the gasoline or diesel engine in your car.

## Heat engines

Some heat engines are internal combustion engines, like in your car, whereas others are external combustion engines, as in a steam engine. The main difference is that an external combustion engine depends on the heat itself—regardless of the fuel—whereas an internal combustion engine is pretty picky about its fuel.

### External combustion engines

There are probably a number of external combustion steam engines around you right now, if you are anywhere near an electric power station. Most large electricity plants use steam engines to turn electric generators that power our homes. That is true regardless of whether they are nuclear, or coal- or gas-fired power plants.

The heat from whichever fuel is used turns water into steam, and then the steam is heated further in an enclosed container, which raises its pressure. The pressurized steam is then fed through a turbine, not unlike a fan blade but much stronger, or a piston engine.

A piston engine can take many forms: both diesel and gasoline engines are piston engines. A piston engine has two major parts, a cylinder and a closely fit piston inside. The hand-pumped air compressor you might've used to fill up the tire of a bicycle in an earlier project is a simple cylinder and piston device. A steam piston engine generally has the pressure built up externally; the steam is then injected into a cylinder and expands to push the piston. Like the piston engines in your car, there are a number of different ways that multiple pistons can be arranged, and a wide variety of innovations have been made over the years.

A turbine engine operates on a principle not unlike your desktop fan, but with air blown through it instead of moving the air forward. The angled blades mounted around an axle slows pressurized steam down as it passes from a high- to lower-pressure containers, and turns the blades backwards, which is often used to turn electric generators. Turbines are of course engineered precisely to maximize the force captured in a way that your fan is not.

Piston steam engines were once the power horses of the railway industry, powered by wood, then coal, and both piston engines and steam turbines are commonly found in power stations around the world. Modern steam-based systems are designed to reuse the steam several times by sending the waste steam from a large turbine or piston engine through a smaller one, and then sometimes a smaller one after that.

External combustion engines have several differences that make them potentially valuable from an energy standpoint. For one, they rely on a temperature difference, rather than a specific fuel type. This can make it easier to combine heat sources, which is useful when you are working with an intermittent heat source, such as the sun. As we saw when installing a solar thermal system, having a backup heat source allows things to continue to function as normal when a renewable source isn't available.

So for example, an external combustion engine could use the sun's heat while it was around, and burn another fuel when it was not.

## Internal combustion engines

Inside the cylinder of a reciprocating gasoline engine, a small amount of fuel is set alight and the explosion that results forces a piston, which fits tightly into the cylinder, away from the explosion. A diesel engine does the same thing but uses pressure, instead of a spark, to ignite the fuel, like we explored in the fire-piston project.

The engine in a vehicle does more than just move the vehicle around, it does several things at the same time: it moves the car forward through gears attached to the wheels, it turns an alternator that generates electricity, and it can send waste heat into the cabin, or turn a compressor to cool the temperature inside the cabin. It makes me wonder, if a car can do so many things at once, why in the basement of so many houses are there water heaters, furnaces, and incoming electricity connections? Couldn't those be combined somehow?

It may not be ideal to have an internal combustion engine (like in our cars) sitting in our basement running all the time, though it is certainly possible. In many places, e.g., remote regions, before centrally provided electricity arrived, internal combustion spark engines did tremendous amounts of work (see Figure 9-1). Engines, like the 100-year old model pictured, are versatile and many were designed to run on multiple fuels depending on what was available. They are, however, a bit loud, and a continuous burn, like you find in a furnace, tends to be cleaner than an intermittent burn, like in an engine.

**Figure 9-1** *An antique gasoline/kerosene-powered factory-type engine, which would have powered machinery and generated electricity before the grid reached these isolated areas. On display in New Brunswick at Home Week in Woodstock.*

# Cleaning up combustion

As mentioned in Chapter 1, burning fuel releases toxic gases into the atmosphere, including greenhouse gases, but depending on the type of fuel, also contributes to acid rain and other problems. Although we aren't quite ready to stop burning everything, we should be as efficient with the fuel we do burn.

Some engines are cleaner than others: e.g., as mentioned, an engine with a continuous burn tends to be cleaner than one with an intermittent burn. A neat way to understand why is to light a match, let it burn for a second, and then blow it out. Pay attention to when smoke is created during this process: it is while the flame is growing, and then from the smoldering matchstick once blown out. The flame itself releases heat, and not much smoke in comparison.

A good general way to reduce pollution from combustion is to increase efficiency by finding more uses for the same energy. Heating systems which also produce electricity are called combined heat and power (CHP) systems. There are a number of ways in which this can be achieved: both steam and sealed-gas (Stirling) engines are contenders that may get more common in the near future; other possibilities are fuel cells and thermoelectronics.

Steam can be a little tricky to work with for the inexperienced, and it requires water heated to over 100°C (212°F). To get those temperatures from the sun we'd have to concentrate it: this is being done, both on a large scale and a scale small enough for us to explore in a later chapter. Below that temperature there is still enough energy to provide work. Not only is a hot air engine a good way to demonstrate that, but also it is being used in commercial heat and power systems that are now coming onto the market for homes and yachts.

Whispergen (Figure 9-2) is one of several companies working on Stirling heat engines for consumer-based

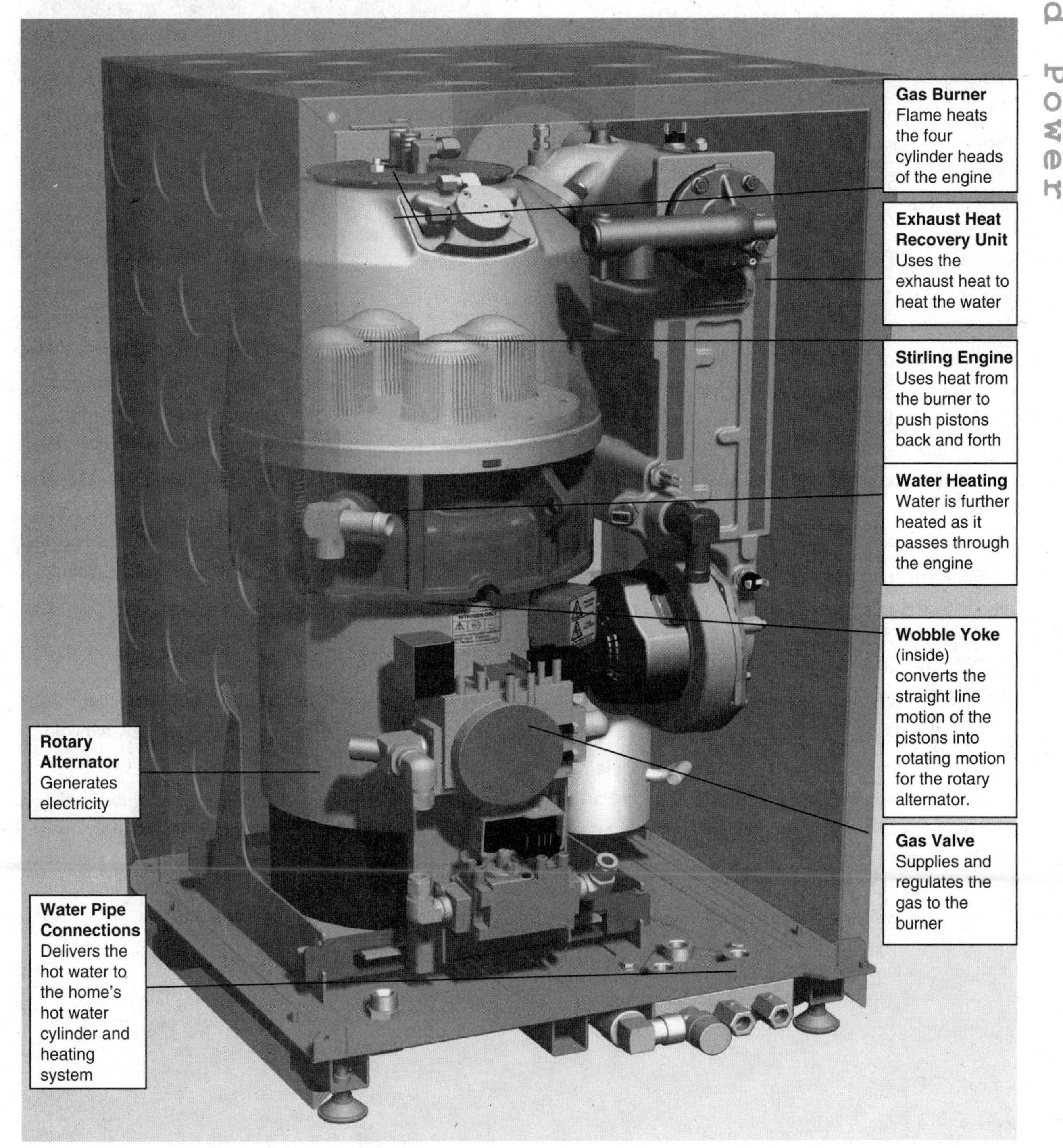


**Figure 9-2** *An internal cutaway of the system behind the Whispergen combined heat and power system, shown mounted in its cabinet earlier. Image supplied and © Whispergen Ltd (www.whispergen.com/).*

CHP systems. The Whispergen uses an existing solid fuel source to generate both heat and power. Other companies have included renewable sources in the mix too, like Sunmachine (see the Online Resources for the link), which uses concentrated solar thermal when available and solid or gaseous fuels otherwise, to power a Stirling-based engine. There are also other companies working on steam version of CHP units, as well as plans for hydrogen or thermoelectric powered systems.

Regardless of the specific technology used, CHP systems increase the efficiency of fuel use because they enable electricity to be produced while fuel is being used for heat.

**Figure 9-3** *The Thinking Man's Engine, available as a kit from PM Research Inc., is a gamma-type engine. It is shown with the optional burner under the hot end of the displacer piston. The separate power piston is seen attached to the far side of the flywheel, which regulates the phase of the piston movements inside the cylinders.*

## Hot air

When a contained volume of air (or any gas) is heated, it will expand; when it is cooled down, it will contract. You can think of the molecules of air as billiard balls: they bounce around very fast when they are heated and slow down and pull close together when cooled. We can turn the energy from this change into useful work using a heat engine.

Among the first to design a heat engine was a man named Stirling who lived in Scotland about 200 years ago. His name is still often attached to machines that only slightly resemble his original engine, because they rely on a principle he first tried to put into practice.

There are three common piston arrangements in a Stirling heat engine. Gamma (Figure 9-3) and beta (Figure 9-4) engines, as the two more common are known, have the power and displacer cylinders either on top of one another (as one cylinder), or as separate cylinders but connected. Inside the cylinder(s) the large displacer piston shuffles a contained volume of air so that it is alternately heated then cooled. The second, usually smaller cylinder contains the power piston, where the expansion and contraction of air pushes or pulls an airtight piston.

The motion of the two pistons is 90° out of phase. This means that when the displacer piston is extended all the way in one direction with the air consequently moved to either all hot or all cold (see Figure 9-5), the power piston is in the middle of its stroke, ready to be acted upon.

The third somewhat less common type of engine is and alpha configuration, where there are hot and cold pistons in separate cylinders.

It is possible to demonstrate these concepts a little more clearly with simple materials, and possibly even create a running engine. You could also order a kit that will run off of a hot cup of water, or even a small burner, all based on the same principle.

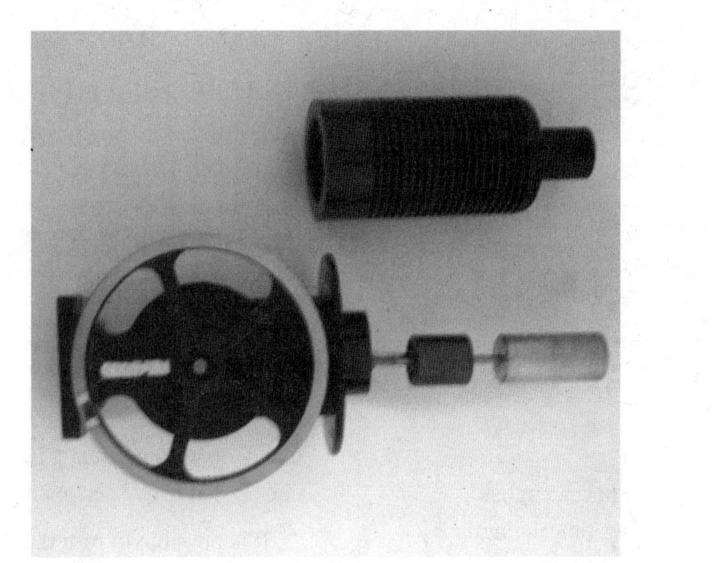

**Figure 9-4** *Beta-type model Sun Runner engine, disassembled to show the displacer (top) and power pistons (bottom) to the right of the cylinder head, which has cooling fins built into it.*

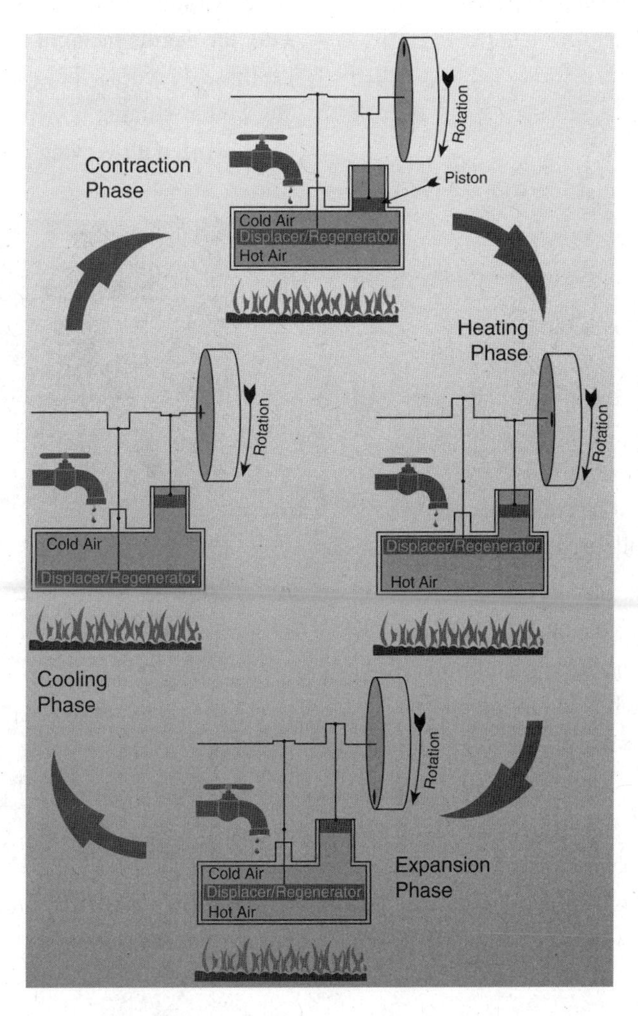


**Figure 9-5** *Diagram of the Stirling cycle through all four phases. Adapted from image courtesy of Brent H Van Arsdell.*

# Project 33: Demonstrating the Power of Hot Air

It is very easy to demonstrate the expansive and cooling forces that the Stirling heat engine takes advantage of.

## Online Resources

- www.bekkoame.ne.jp/~khirata/english/mk_can.htm—the original tin can engine by Koichi Hirata, versions of which have been built around the world and instructions and modifications posted all over the net.
- members.aol.com/hstierhof/index.html—Hubert Stierhof has done some innovative work on useful but simple engines based on hot air and a moving cylinder.

## You will need

- A tin can.
- A balloon that fits tightly over the top.

## Steps

1. Take a small metal can that has one end completely removed, and stretch a balloon tight across the open end.
2. Put a heat source under the can and the balloon will start to inflate. The tin can acts as the heated displacer cylinder and the balloon as the power piston, with the force of expansion acting to inflate it or push it outwards.
3. If you then place the hot can (carefully) into a cold bowl of water, the balloon shrinks from its inflated state and, if the water is cold enough, sucks the balloon into the can.

The up-and-down motion can then be used for useful things. It could turn a wheel, or do a job that required an up-and-down motion, like power a pump. Both motions are useful for, among many other things, producing electricity. Of course it is not that convenient to be moving the can from the flame to a cold water bath, so these two temperatures are usually held separate in practice, while the air is moved back and forth between them.

It is possible to turn the demonstration tin can and balloon into an engine of sorts. To do this, we reduce the amount of air inside by a half, by displacing the other half with a lightweight piston. In instructions on a project along these lines, Koichi Hirata suggests using balsa wood as a displacer piston. Students at the Centre for Alternative Technology, Machynlleth, Wales, UK, built the engine pictured in Figure 9-6, relying roughly on those directions, but used other displacer materials (as well as coming up with the condom alternative on their own), with varying results.

**Figure 9-6** *The model tin can engine built by students at the Centre for Alternative Technology.*

With a displacer roughly half the size of the cylinder, which is a loose fit, acting as a displacer piston, we can design the top half of the tin can as the cold side of the engine. Because the heat will be further away from it (relative to the hot side), it will be colder than the hot side. The displacer is expected to move up and down in the tin can, with the flame at the bottom (see Figure 9-7)

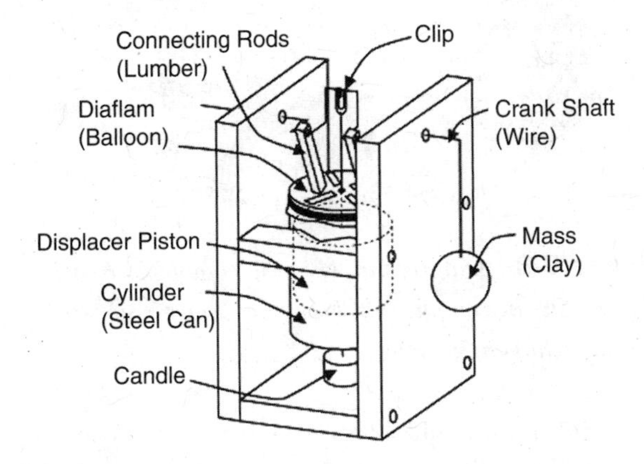


**Figure 9-7** *A diagram explaining the major parts of the tin can engine you can try to build.*

## You will need

- A fishing line.
- A wooden base and walls.
- A wire coat hanger.
- A displacer cylinder made from balsa wood or improvised (but be careful) from a lightweight material.
- A tea light or other heat source.
- A drop of oil.

## Steps

1. Attach a small piece of fishing line to the top of the displacer, and fashion an axle with a 90° phase, from an old coat hanger. Search for alternative designs online if your materials don't match exactly, and be creative.
2. Fix the axle to an improvised stand.
3. Attach the other end of the fishing line to the axle, first feeding it through a very tiny hole in the balloon that is poked dead center.
4. Tape another piece of fishing line to the top of the balloon. On the end of the axle, either place a flywheel, or a 90° bend with a counterweight.
5. Put a small drop of oil on the fishing line that passes through the balloon to seal the hole.
6. Place a tea light under the can, and rotate the flywheel (or weighted axle) to cycle the engine. This engine does not start itself. Once the air inside is warm enough, and if the air in the room is cool enough, spinning the axle should start the engine turning. The directions must be followed precisely because the frictional forces are difficult to overcome.

# Project 34: The Energy of a Hot Cup of Tea

The Eco-Power Stirling Engine Kit from Stirlingengine.com is designed to run off of a hot cup of tea. Order the kit along with the book and homopolar motor kit (a project elsewhere in this book) to save on shipping costs. It does take some work, but it can also be very educational. There are alternative engines listed on Evilgeniusonline.com, which will even run off the heat of your hand.

## You will need

- The kit from Stirlingengine.com.
- A few items in addition to the kit, including:
  - a Hammer
  - 1/16 inch drill bit
  - epoxy
  - white glue
  - cyanocrylate glue
  - double-sided tape
  - a mug.

## Notes

1. Simplified directions are difficult for this project, but we'll give you a good idea of what to expect if you do decide to order the kit. A diagram of the completed engine (Figure 9-8) compared to the parts that come with the kit (Figure 9-9) gives the reader an idea of the attention to detail that will be required.
2. There is a video that comes with the kit and is well worth watching. You probably want to watch the whole thing through once or twice before you start: it can get complicated, particularly if you are not sure of what the next step will be.

## What's what

- The displacer cylinder (far left mid-way up in Figure 9-9) is thin plastic and the piston expanded foam (next right). The hot and cold ends of the engine are two pieces of cut aluminum (top left), which you will bend into the correct shape. The power piston is, as in the last engine, a piece of stretchy rubber and the axle is again a thin piece of wire.
- Most of the body and the flywheel are pre-cut cardboard pieces held together with glue and epoxy. After popping the parts out of the cardboard sheets, the included instructions show you exactly where to fold and tape (Figure 9-10) to make the body.

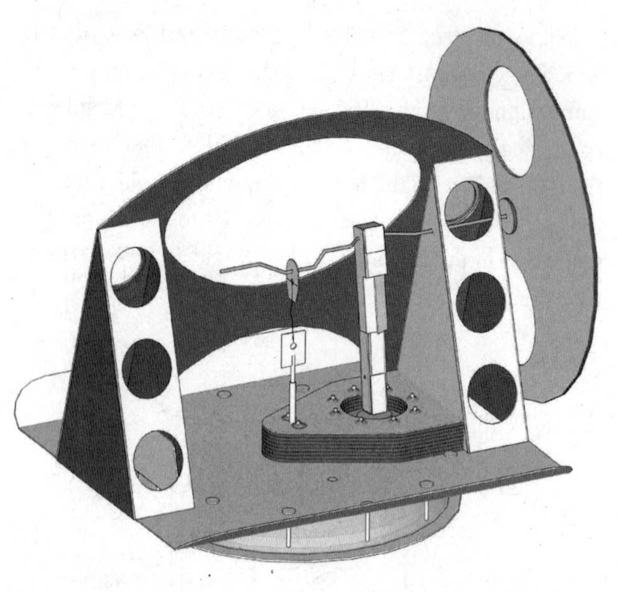

**Figure 9-8** *Diagram of what the completed Eco-Power Stirling engine will look like. Image courtesy of www.stirlingengine.com.*

- The kit then guides you through seven easy-to-understand steps, including assembly of the connecting rod, diaphragm (Figure 9-11), cylinder, and displacer, then mounting the bottom plate, superstructure, crankshaft, and flywheel. This is followed by some testing and troubleshooting tips: as with most Stirling engine models, these include techniques on tracking down leaks and sources of friction in the engine.

**Figure 9-9** *Parts that come with the Eco-Power Stirling Kit, from American Stirling. Image courtesy of www.stirlingengine.com.*

**Figure 9-10** *One of the diagrams from the Eco-Power Stirling Engine kit shows how to fold the superstructure. Image courtesy of www.stirlingengine.com.*

## Heat pumps

Heat pumps and air conditioners are heat engines working in reverse: i.e., they draw energy from another source and use that energy to do work on a temperature difference. They often employ evaporation and compression cycles of a very specialized working fluid that has convenient vaporization and condensation temperatures. It is one of these working fluids that has been the cause of a problem many people have heard of: the release of CFCs (chlorofluorocarbons) from older refrigerators and air-conditioning units. Thought to be benign because they didn't seem to react with anything for many years, CFCs have been implicated in destroying the ozone layer and, precisely because they are fairly unreactive, they take a long time to clear up from the atmosphere. Though very few are released anymore, those that have already been released will be in the atmosphere well into the next century.

Thankfully there are alternatives, and they are now coming onto the market. The simple heat engines we examined in this chapter have shown that a small temperature difference can be turned into useable work, and we'll soon see in more detail how cold can be created by a Stirling engine. Cold can be created in many different ways. It is likely the reader who has taken a long car trip with a cooler that plugs into their electric socket is aware of another type of cooling device that also has few harmful chemicals used inside it: a thermocouple. The thermoelectric devices inside DC fridges are very robust cooling devices because they have no moving parts. They rely on different types of thermocouple material to generate a temperature difference and they can do the opposite—generate electricity from a temperature difference—too. They are typically designed for a specific temperature range and have some pretty precise manufacturing to them, so they are not that easy to play with. However, their function is similar to the heat engines in the projects.

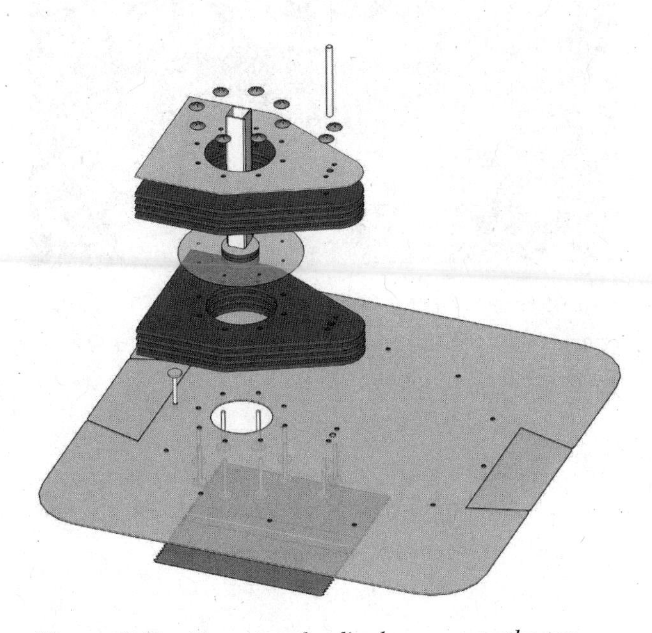

**Figure 9-11** *Mounting the diaphragm onto the top plate. Image courtesy of www.stirlingengine.com.*

## Solar electric options

There are two common ways to use solar power to create electricity. The first and oldest is to use the heat of the sun in a heat engine of some sort. Early experiments tried steam and air engines with moderate success. One of the challenges is getting solar energy concentrated enough to be useful to heat engines, which tend to prefer high-grade heat—around 400°C (752°F). It is possible to get solar energy concentrated to this extent but high-quality materials are needed for economical performance.

There are several ways to go about concentrating solar energy to raise the temperature at a collector point: not all of them need be complicated. The idea is generally to take the sunlight hitting a large area and direct it onto a smaller area. This can be done with lenses, as Gavin Harper showed in *Solar Energy Projects*, or with mirrors placed and aligned on the ground pointed at a single higher point. Other systems are successful with parabolas or even just semi-circular pieces of reflective material. There are designs available for concentrating dishes made out of pop cans (see Evilgeniusonline.com), as well as sites that can help you calculate the shapes of your dish. It is possible to reuse old materials, like a discarded satellite dish or soda cans. You can experiment with a couple of different mirrors and bouncing the light that strikes them onto a single point, and measure the temperature there—which can get a lot hotter than the 80°C or so collected as water earlier!

Concentrated solar energy can easily reach 400°C (752°F) and higher, so do be careful. Large-scale solar thermal plants exist around the world, powering both steam and Stirling engines that feed electricity onto the grid.

## Project 35: Concentrated Solar Thermal Motion

This project would need the enthusiastic evil genius to hunt down the Sun Runner Stirling Engine Kit, or an equivalent. The concentrator (Figure 9-12) is available from www.pmresearchinc.com and success can be had using other engines such as the Thinking Man's Engine, shown earlier and still available from PM Research Inc., with the concentrator. The sun's energy is concentrated on the very tip of the engine and heats up the air inside the cylinder, which expands against the power piston, causing it (by turning the flywheel) to move the displacer cylinder. This shifts the engine into the cold cycle, where the same air that was just heated is cooled, and the compressive force it exerts pulls the power piston back into the cylinder, moving the displacer again and returning the engine to the heating cycle. The concentrated solar heat could, in theory, be used to turn water into pressurized steam that could be sent through a piston engine, like the model one shown in Figure 9-13, available from PM Research Inc.

There are very large systems of this sort, but I have not come across small-scale experiments, yet.

**Figure 9-12** *The Sun Runner Engine, made by solar engines with a solar concentrator sold through PM Research Inc., in action.*

**Figure 9-13** *The Model #4CI working vertical steam engine, available from PM Research Inc. Image courtesy of PM Research Inc.*

## Online Resources

- www.pmresearchinc.com is a wonderful retailer of all sorts of fantastic model engines, including the Sunmachine Engine used in this project.
- www.baileycraft.com is the original manufacturer of the Sunrunner engine and concentrator dish. They also make a variety of other interesting and educational engines.
- www.newenergyshop.com/ sells a direct sunlight demonstration Stirling engine and a variety of other very detailed experimentation kits.
- www.sunmachine.com/ takes the idea behind the Sun Runner model engine and has created a home-sized solar or biomass co-fired concentrating heat and power system around their Stirling engine technology.

## You will need

- The Sun Runner or a similar heat engine.
- Solar concentrator.
- An electric generator that can be attached to the engine or an alternative way of measuring the power produced by the engine can be devised. An interesting low-tech way, shown in the images, is to use a breaking dynamometer constructed from:
  - a piece of a cotton string
  - a container attached to the string
  - a hanging scale
  - several equal weights (coins work well).
- To operate the dynamometer, the cord is attached to the scale on one end and wrapped once around the moving axle (Figure 9-14), such that the motion of the axle will lift the container that is attached to the other end. Known weights are gradually placed in the container (Figure 9-15), and the apparent weight recorded on the scale is used with the actual weight of the coins and the speed of the axle's rotation

**Figure 9-14** *Scale and cotton cord wrapped around the axle of the solar Stirling engine, acting as a breaking dynamometer. Note that for accuracy, the scale should be held securely in place and a more accurate scale could be used.*

Figure 9-15 *Add a coin to the basket and record the apparent weight while the engine is running. Using a high-speed video camera (not shown), or other device to record the speed, will enable you to calculate the power produced by the engine under different loads.*

(counting revolutions from a video of the wheel spinning works), to estimate a power curve for the engine. You can use this to size an appropriate electric generator and load for a more permanent installation.

- A load or means of measuring the electricity produced.
- Optional: a method of drawing waste heat away from the cold side of the engine: be creative and explore different options (see below).

## Steps

1. The Sun Runner Engine, like most Stirling engines, requires almost no maintenance. If you did just want to see how the inside worked (shown in Figure 9-4), you'll want to carefully slide the pistons back into the cylinder (Figure 9-16).
2. Set up the engine on a tripod and, if you forgot, tighten the small Allen bolt (Figure 9-17) on the side to hold the cylinder and heat sink onto the base.
3. Point the dish at the sun and align it so that a white strip is clearly seen across the tip of the engine.
4. Use one finger to gently spin the flywheel and start the engine turning (Figure 9-18).

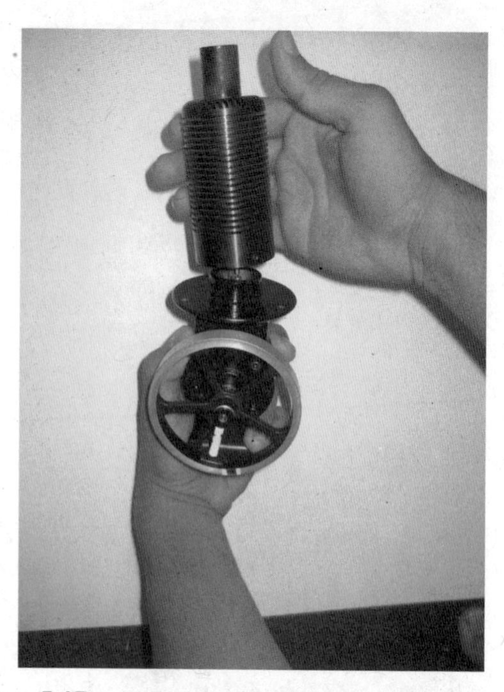

Figure 9-16 *Sliding the cylinder and cooling fins onto the body of the Sun Runner model Stirling engine. concentrator dish.*

5. Once it is turning, attach the generator and load, or attach a dynamometer to measure the torque produced under simulated loads. You might also measure the temperature attained from just the sun's heat (Figure 9-19).
6. Marvel at one 100-year-old solar power!

Figure 9-17 *Tightening the tiny Allen bolts on the side of the cylinder body that hold it in place and keep a tight seal.*

**Figure 9-18** *Use one finger to start the engine turning. On a larger system, an automatic starter would of course serve this function.*

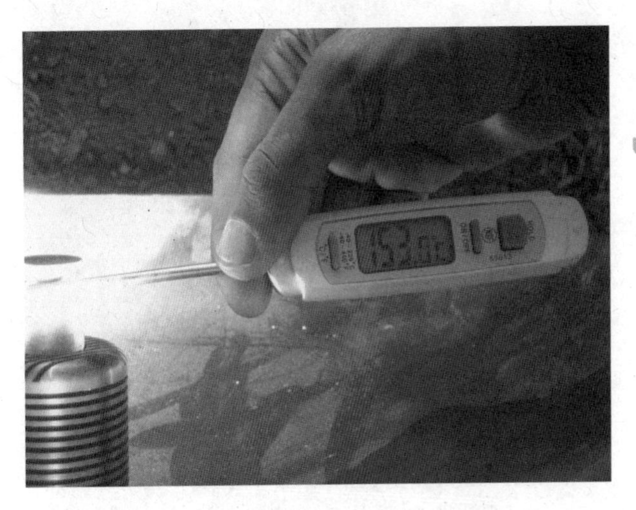

**Figure 9-19** *Use a simple oven thermometer to measure the approximate temperature at the apex of the concentrator dish.*

## Variation

Cool the cold side of the engine and try to increase the amount of power generated. The engine will function without extra cooling, using the fins on the cold side of the cylinder for cooling. To help more heat escape, try blowing more air past the cold side of the engine with a fan.

For a more advanced project, run copper tubes with cold water in them around the cold side of the engine, into a bucket of water. If the bucket is below the engine, a thermosiphon effect will draw the cold water up and return heated water. As we spend so much of our fuel heating water already, isn't this a great by-product of solar electricity? Take a second to compare the power produced before and after adding a cooling system: it should have increased.

# Project 36: A Thinking Man's Cold

Low-temperature engines, like those in the first two projects of this chapter, are good demonstration tools of the energy that is available to be put to use from even very small temperature differences. The Thinking Man's Engine (available from PM Research Inc. and shown in Figure 9-3), one of several different models available from hobby-sized builders around the world that operate at a higher temperature difference, is more precisely machined. Some suppliers are listed on Evilgeniusonline.com but the list is not exhaustive. Look around and try to pick one made close by if you can find one. The project definitely isn't specific to the engine, but it is something that perhaps not every engine owner has done yet. The accompanying images show the Sun Runner Engine from the last project and the water wheel from Chapter 8, although students in class at the Centre for Alternative Technology in Machynlleth, Wales inspired the project using the Thinking Man's Engine and an electric drill, with more success.

## You will need

- A medium-temperature heat engine.
- A small electric engine or other mechanical rotary power source (a second heat engine driven by a heat source will suffice, as will the water wheel from a project in Chapter 8 if you can wait that long!).
- A belt or alternative.
- Possibly a pulley, particularly if the driving engine is not variable speed.

## Steps

1. Ensure that the engine is securely held in place (one set of hands will do, but make sure there is another set around to hold the other engine).
2. If using the Thinking Man's Engine or an engine with a similar feature, you can remove the flywheel and attach the belt drive.
3. Fit the belt around the drive shaft of the Stirling engine and around the other power source, so the belt is tight.
4. Turn the stirling heat engine backwards using the second power source (Figure 9-20).
5. Create cold (Figure 9-21)!

**Figure 9-20** *The Sun Runner Stirling engine, attached by a thin belt and powered by the model water wheel from Chapter 8.*

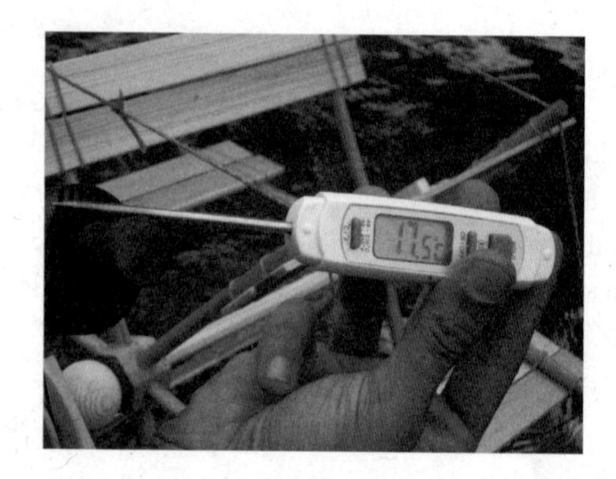

**Figure 9-21** *A small but measurable temperature difference was created, though it did take some time!*

## Efficient or convenient

One of the reasons why CFCs became so popular more than 50 years ago and heralded a new age of refrigeration for all, though there had been refrigerators before, is that CFCs were efficient at the job of moving heat from one place to another. They were also relatively inexpensive to produce and relatively safe for the user.

Now that we are looking around for the ability to reduce the use of harmful substances, it is comforting to know that it is possible with a correctly designed heat engine using nothing but air as a working fluid. However, if we do the math of the efficiency of the temperature difference produced in our earlier project, it is not good. This is a large part of why substances that have deleterious effects on the planet continue to be in use, because the alternative often does not have the technological maturity to compete commercially. The fundamental technology behind a Stirling cycle—a contained gas heated then cooled—is already widely used in the cryogenic freezing industry because it is an efficient way to create such a low temperature and has been producing electricity in space. The potential in the technology is there, and the inspiration for a designer to build an efficient consumer-grade version may just be waiting for us to want it.

Heat pumps operate on a similar principle, but they, like air conditioners, have a greater efficiency than the devices we have demonstrated so far. Heat pumps concentrate (or move) heat using another power source, such as an electric motor, and use an air or underground heat exchanger as an initial stable temperature. They also work best in particular temperature ranges, and mild climate tends to be best. Ground source heat systems often rely on the Earth remaining at a constant temperature below a certain depth, regardless of the season in most parts of the planet. A geothermal system, on the other hand, will depend on a relatively hot piece of ground that occurs naturally in only a few different places, for most of their heat, possibly using a heat pump to raise the temperature during part of the season but also potentially producing power when there is spare heat that is not needed. A ground source heat pump with no geothermal source will never do such a thing, but as we've seen, the ground can also be used to store heat. During times of plenty of heat, like the summer, developments like Drake Landing outside of Calgary, Alberta (www.dlsc.ca/) are storing solar heat through the seasons, and drawing it out during the winter when it is most needed. Genius, eh?

# Chapter 10

# Electricity

We have been using electricity for a long time. From the days of Thomas Edison, to the world of electronics engineering today, researchers continue to discover new and interesting characteristics of electricity. We'll discuss recent advances to generating electricity in later chapters, but for now we'll focus on more "traditional" ways of creating and storing electricity, which have changed only slightly in over 100 years.

## The basics

Electricity is the flow of charge through a conductor. Basically, the world is made up of atoms. Each atom is made up of protons, neutrons, and electrons. Protons and electrons are positively and negatively charged, respectively, and neutrons have no charge at all. In an atom, protons and neutrons are large and bound tightly together, whereas electrons are small and bound loosely to the neutrons and protons. Most particles have no net charge because the number of protons in the atom equals the number of electrons. A particle becomes charged when electrons move from one atom to another. For example if two neutral particles get close together and electrons move from one to the other, then one particle becomes positively charged and the other negatively charged. If you walk along a carpet in your socks and see little sparks, this is particles becoming charged, or atoms taking on and loosing extra electrons. Now if we take such a charge and move it quickly through a conductor—that is a material made up of particles with loosely bound electrons, like a copper wire—we get electricity. The movement of electrons is known as a current—it can be described as a flow, in the way that a liquid flows, of electrons.

Voltage is the measure of electric potential. A common analogy is to a faucet of water on a bucket, where the voltage (in volts, abbreviated V) is analogous to how wide the faucet is open, and the current (measured in amps, abbreviated A) to how fast the water is flowing through that opening. The combination of the voltage and current gives an idea of the power of the stream of water. Multiplying the voltage by current results in a measure of energy use over time, known as power, measured in watts (W).

There are two types of current of electricity. Though we are not always aware of it, most people use a mixture of alternating current (AC) and direct current (DC) throughout their day. The light sockets in most houses provide AC from the power company, whereas your car battery and cell phone both use DC. The difference is that alternating current, as its name suggests, alternates the direction of current flow, typically many times per second. So while a battery has one positive terminal and another negative terminal, which makes it pretty easy to imagine electrons traveling from one to the other down a copper wire, AC gets a bit more difficult to imagine. If you had a chance to look at an oscilloscope of some AC current, you would see a sine wave traveling across the screen. The AC current coming out of the sockets in your house alternates between 50 and 60 times per second, meaning that the sine wave touches the positive or negative extremities at least every 1/100th of a second!

### Generators and motors

A common way of creating electricity is to use a spinning power source, like a shaft, to turn coils of copper wire in between a magnetic field, often created by a pair of permanent magnets (see Figure 10-1 for a close-up of small version), alternately using an electromagnet—created by passing electrons through a second series of copper coils. When the coil spins, it creates a current.

**Figure 10-1** *A model electric generator. The coil of copper wire can be clearly seen on the axis (vertical down the image) and the magnets (encased in plastic) on either side.*

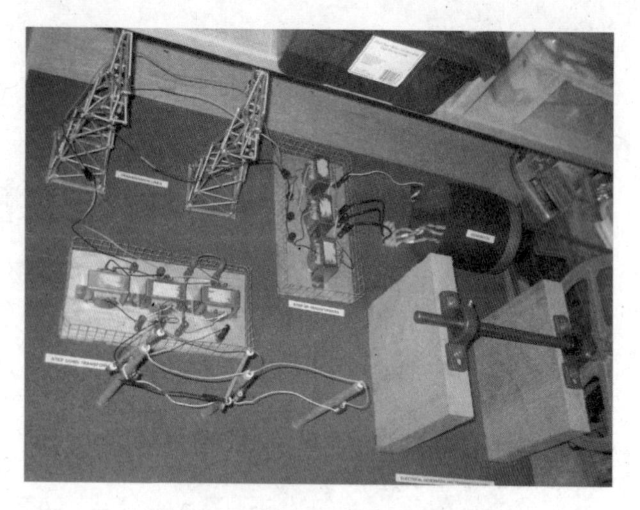

**Figure 10-2** *A demonstration of the electric grid: a generator (top right) connects to a step-up transformer (center), feeding into high-voltage power (transmission) lines (top left), then stepped-down to power the street lights (bottom left).*

An electric generator will produce different levels of current and voltage, depending on how it is wired and how fast it is turned, and an electric motor will similarly be "tuned" to a particular power supply. For electricity to be useful, the voltage and current need to be well matched to the application or load on the circuit. For example, those who have crossed the Atlantic Ocean will know that the voltage used on either side is different, and that their appliances need to be able to accept different voltages, or have something that will adapt the voltage to their appliance, from the 220–230 V range to the 110–120 V range.

Different voltages of electricity have different properties and characteristics and, as a result, uses and locations where they are most appropriate. For example, very high voltages are used by electricity companies to transmit power over large distances, because less electricity is lost in transmitting high-voltage electricity than low-voltage electricity, through the same cables. Figure 10-2 shows the basic parts of a typical part of the grid: the hand-turned generator (not common on a large scale) feeds electricity to a step-up transformer, which feeds high-voltage electricity into the power lines, which is then sent through a step-down transformer—to turn very high voltage electricity into something that the light bulbs can use—before powering the street lights.

Medium-sized electric generators are found in places like the alternator of your car, which uses the force created by the engine running to recharge the battery of your car, and at the center of a wind turbine to turn the spinning motion of the blades into useful electricity. One of the neat things about electricity is that the same combination of magnets and coils of copper wire are found in many electric motors as well. Though the arrangement and number of coils of copper wire would be slightly different in something designed to be a generator compared to a motor, the concept is the same and many motors can be used as generators.

Simple demonstration kits can be found at a variety of stores; probably the local hobby shop is your best bet to keep the transportation emissions of your projects to a minimum (see Figure 10-3 for an idea of what to look for). If you are the adventurous type, you could take apart an old motor lying around to get a better understanding of how motors and generators work. I always found that hands-on tearing something apart is the best way to learn. You might find an old motor in a tape deck or eight-track player in your basement, but make sure no one will miss it before you take it apart! You could then attempt to attach it (via a belt or by direct drive) to some of the prime movers used in our later projects, and experiment with different sizes of electric generators. Another interesting way to get an understanding of how much work must go into providing even a small electric charge is to use, even just in an emergency, a hand- or pedal-powered electric device.

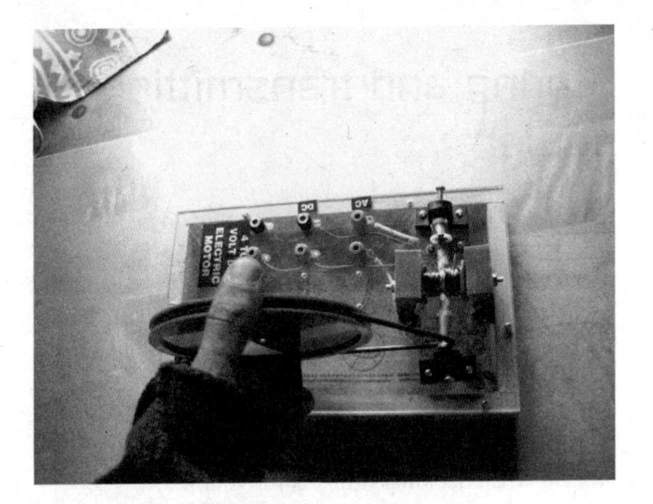


**Figure 10-3** *The layout of a demonstration electric generator kit: a large wheel is hand-turned to drive an electric generator, creating current at the terminals.*

### Online Resources

- www.magnet.fsu.edu/education/tutorials/java/faradaymotor/index.html—a good tutorial for those wishing to understand more about how this motor works.
- www.evilmadscientist.com/article.php/HomopolarMotor—Evil Mad Scientist Laboratories really do "make the world a better place, one mad scientist at a time."
- www.eia.doe.gov/kids/energyfacts/sources/electricity.html—the US Department of Energy has put together a wonderful refresher course, for anyone who isn't currently in a physics class.
- ocw.mit.edu/OcwWeb/Physics/8-02Electricity-and-MagnetismSpring2002/CourseHome/—The open courseware site of MIT is a treasure trove of useful information on a range of topics.

## Project 37: The Homopolar Motor

The homopolar motor is a very simple motor that can help the reader to understand how electricity works. The concept behind it dates back to a device created in the 1800s and credited to Michael Faraday. The homopolar (or Faraday) motor does work spinning a screw using the charge created by a battery and the magnetic forces created by the magnet; it demonstrates how electric and magnetic forces can be used to create rotational movement.

### You will need

- Order the kit from American Stirling to save yourself the trouble of hunting down the hard-to-find bits of parts; otherwise, look up some recommendations of sources using the Online Resources provided.
- The materials aren't very enormous really (see Figure 10-4). One AA battery, some copper wire, a screw, and a really special magnet (such as a neodymium disk magnet).

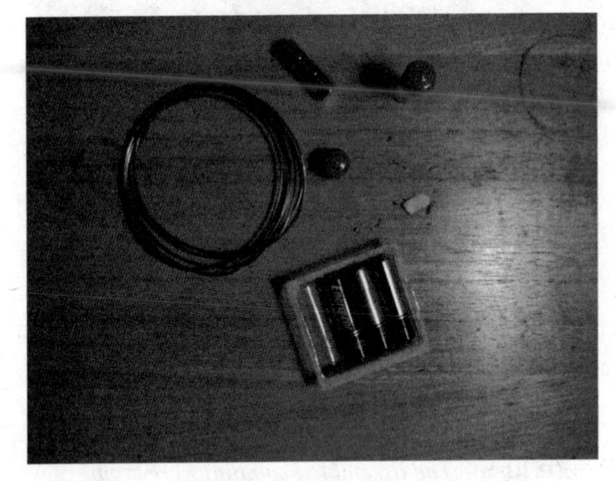

**Figure 10-4** *Parts that arrive with the American Stirling kit (www.americanstirling.com).*

## Steps

1. There are two easy ways to build a homopolar motor. The American Stirling kit suggests bending the copper wire so that it secures the magnet directly to the base of the battery, causing the magnet to spin. In practice, we have found that adding a common household screw to the motor provides a more satisfying demonstration.
2. Place the magnet on the base of the screw and the other end hanging (using the magnetic force acting through the screw—notice that there is a maximum size of screw and magnet that can be used, else the screw will not be held to the battery with the magnetic force) from the negative end of the battery.
3. Touch the copper wire to the positive battery terminal.
4. Lightly touch the opposite end of the copper wire to the end of the magnet without the screw hanging from it (see Figure 10-5).
5. The screw will start spinning. If not, check that there is a good contact at both ends of the copper wire.

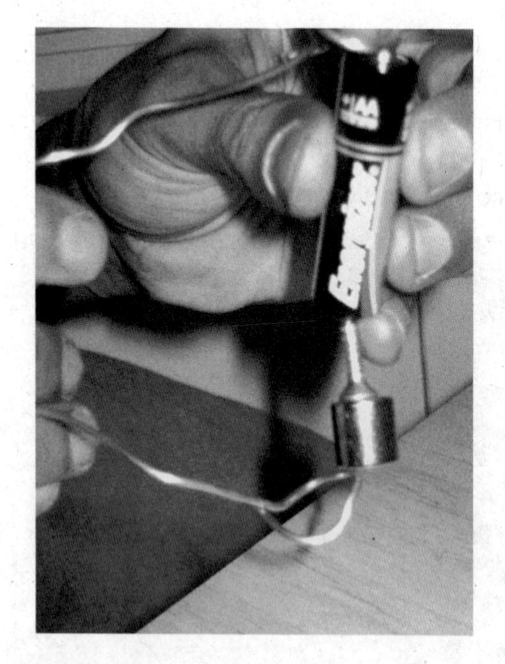


**Figure 10-5** *The assembled spinning screw-type homopolar engine.*

## Storing and transmitting power

Because renewable sources of power tend to be intermittent, energy storage quickly becomes an issue for those wishing to use the sun or the wind as their primary source of electricity. It is possible to use a battery to store some electricity, but as the most common and affordable chemical batteries have a not inconsiderable environmental impact, it is good to be prudent when deciding if this is for you.

Chemical batteries are formed by the interaction of two different chemicals. Some batteries are consumed by the process—known as single-use batteries such as common alkaline batteries—whereas others can accept and deliver a charge many times—such as the lead-acid battery in most cars—over their life. Lead-acid batteries are common and relatively affordable. There is also a well-developed system for recycling both the lead and the acid in many countries around the world. Most batteries have a certain lifetime: lead-acid batteries commonly last 5 years with good maintenance, whereas nickel-metal hydride (NiMH) batteries and other advanced batteries can last longer with proper care. At the end of their life, all chemical batteries need to be disposed of with care to the appropriate facility that can accept and handle the materials inside safely. Energy goes into all phases of chemical battery manufacturing. In many cases this can be quite a lot of energy, so choose batteries carefully and care for them well if you must use them. There are advanced benign alternatives, like fuel cells, which can also function as batteries, but there are challenges to commercialization (such as storing hydrogen safely) that prevent widespread adoption at the moment. Also, consider some of the more benign alternatives just covered in the last chapters—air and water—depending on the landscape around you.

If you are somewhere remote that has no other electricity supply, then installing and using chemical batteries might be the only option. In this case it becomes important to consider the size of the batteries and how much power is going to be required, for how long, when, and how critically. Battery systems typically require some trade-offs to reach a compromise between useability and cost.

Many people who have traveled in a motor home or camped with a trailer that had its own battery system may have experienced what a conservation lifestyle entails. Batteries only store a certain amount of energy, and they have a funny way of delivering it. If you draw a lot of power from a battery quickly, you will find that overall you get less energy back from it, than if you drew a small amount of power out over a smaller period of time. Batteries also only hold so much energy, so they require the people using them for necessities to be somewhat considerate of their energy requirements. Getting people to be conscious about their electricity consumption is useful for the environment, even if batteries are not always very good. Knowing that, it is a good idea to take account of your energy use, in much the same way as a person who was considering buying a battery-based electricity system for their home would think about what appliances they absolutely could not live without.



## Project 38: A Coin Battery

It is common for us to have batteries of different sorts all around us and not really understand that part of what is happening is a chemical reaction between two metals. This super-simple battery is potentially useful to get an idea of a very simple device that produces a small electrical charge. We use really benign materials to make it (Figure 10-6) but that shouldn't detract from the reality that consumer-grade batteries often contain dangerous chemicals and should always be disposed of properly.

A battery is typically composed of cells wired together. In this project, the copper parts of the coin react with the aluminum sheets in the presence of an alkaline solution, which is soaked into layers placed between the copper and aluminum. Increasing the number of cells (i.e., coins) stacked in order will increase the voltage of the battery.

### Online Resources

- www.users.globalnet.co.uk/~jchap/tvproba.html.
- brightest-kidz.squarespace.com/journal/2008/1/17/coin-powered-battery-project.html.
- www.sciencecube.com/Eng/education/mbl/mbl-coinemf1.asp.

### You will need

- Six pennies or copper-coated coins. Pennies are reported to work in the United States, Canada, and the UK, and 50 paisa coins (among others) in India.
- A small dish.

**Figure 10-6** *Simple materials for this project: a dish, coins, aluminum foil, wire, and salty water.*

- Some aluminum foil (enough to trace out six pennies). Another alternative is to use 10 cent pieces if you are in the United States or Canada, as they have some aluminum in them that can be a substitute. It is possible you can find other coins or materials as a substitute; it is important that they have some aluminum, as this will react with the copper in the saline solution.
- Scissors to cut the aluminum foil (if you are using it).
- Warm salty water (another alternative is lemon juice for a more potent battery).
- Two small pieces of wire and some tape or solder and a soldering iron.
- Paper kitchen towel or pieces of cardboard or paper to hold some saltwater or juice in between parts of the cells.
- A load (a small buzzer or LED light could work) and/or a multimeter.

## Steps

1. Cut out circles of aluminum foil and paper kitchen towel as big as the pennies (Figure 10-7).
2. Tape the bare end of one piece of the wire in to one penny, and the bare end of another wire to a foil circle. These will be the positive and negative terminals of the battery.
3. Moisten one small paper towel with the salty water.
4. Place the foil circle with a wire in your dish and place one moistened piece of paper towel over the top (Figure 10-8).

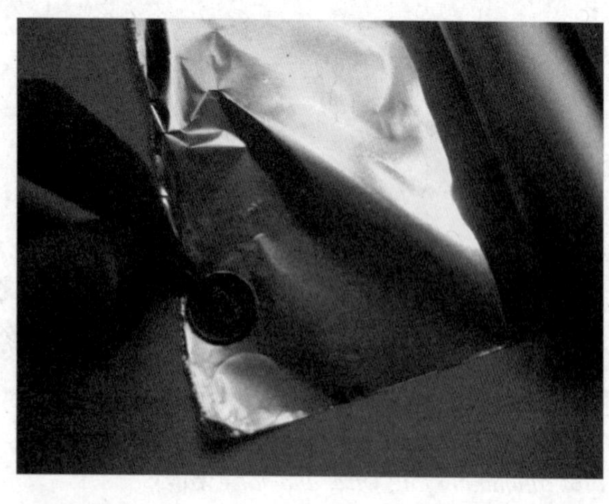

Figure 10-7 *Cutting penny-sized circles from tin foil.*

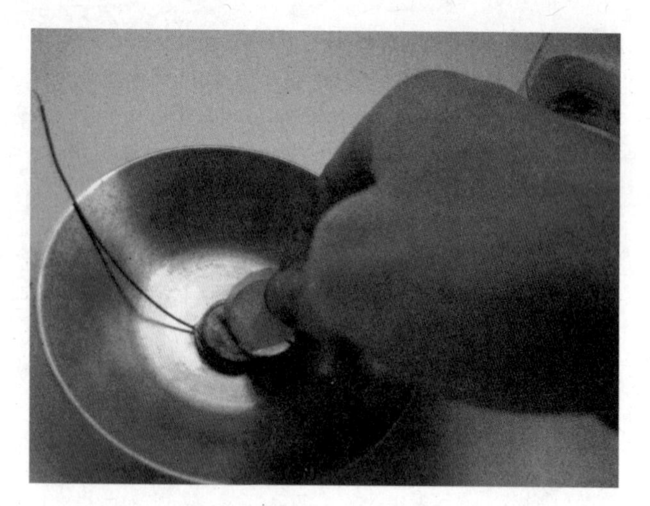

Figure 10-8 *Layering coins, aluminum, and wet paper towel to form a battery.*

Figure 10-9 *The completed battery.*

5. Cover the paper towel with a penny, and the penny with a circle of tin foil. Repeat this layering; your top coin should have the wire taped to it (Figure 10-9).
6. Attach a (very small) load or multimeter to demonstrate your battery.

## Moving it around

If you are not going to store electricity, then it either needs to be used or put somewhere where it can be used. The grid is a system of transmission and distribution that crosses the continent and covers many parts of the world.

In some places, where allowed by the utilities or body that controls the power grid, it is possible to act like a generator and feed power back onto the grid. In order to do that, the energy being put back must exactly match the grid; otherwise, lots of trouble could happen. Thankfully, there are some very good products that now enable everyday people like you and me to generate clean renewable electricity and sell it back to our power company (if they are willing) for our neighbor to use. You will need to have specialized equipment that interacts with the grid and which your electricity company will allow. Whether it is worth it for you to consider buying this sort of equipment very often depends on how much your power company is willing to pay for the electricity you are generating.

The most generous places will pay a premium for types of electricity they want to encourage, such as solar- or wind-generated electricity. Germany and Ontario pay very high rates for solar electricity produced and sold onto the grid in order to encourage customers to become generators as well as consumers. A middle of the road approach would allow the customer to rollback their consumption with production, which is called net metering. Net metering can have the advantage of increasing awareness by consumers of the amount of electricity they are using compared to what sort of resources are necessary to provide that electricity. The vast majority of electricity markets remain without real alternatives for customers to interact with the grid. This is a policy problem and, as we discuss elsewhere, getting active and making your voice heard to politicians and industry representatives is a very green evil genius project in itself. So do some reading, then add "net metering" to your lexicon and start pounding the pavement!

## Charge controllers and inverters

In setting up renewable energy systems, electricity will need to be managed in order to move it from generators to storage devices or the grid. Charge controllers and inverters will likely form a part of most renewable energy systems.

Charge controllers deliver the appropriate amount of electricity, depending on the battery's state of charge and type of battery. They will often be specific to more advanced batteries. Lead-acid batteries can absorb a lot of charge when they are between half and mostly full,

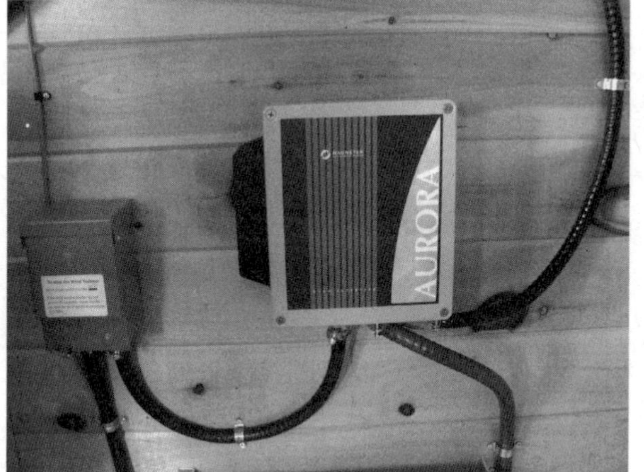


**Figure 10-10** *An Aurora grid-tied charge controller that is specific to the turbine shown in Figure 10-11 is connected to a net-metered system in New Brunswick, Canada.*

but like to be filled up slowly at the end. A smart charge controller will know this, and many other things about your battery, and often about the source of electricity you are charging it with. The controller pictured in Figure 10-10, for example, is specific to the turbine (Figure 10-11), and monitors windspeed so it can rectify

**Figure 10-11** *The Aurora turbine tilting out of the wind to protect it from spinning too fast and damaging the turbine or electronics.*

**Figure 10-12** *A solar- and wind-powered net-metered electric system at the Falls Brook Centre in New Brunswick. Bottom right is the dump load, in case too much power is being produced, because there is no battery. Above the dump load is the Aurora wind controller, left of it is an emergency shut-off switch, and further to the left are the meter and junction box where the solar controller (top left, out of picture) is connected.*

**Figure 10-13** *Flooded lead-acid batteries on the floor of this educational setup at the Falls Brook Centre, New Brunswick. The small charge controller and fuse box are mounted on the wall above.*

**Figure 10-14** *The door of the fuse box of the above system, open while attaching the solar electric panels from Chapter 11. Brent Crowhurst from the Falls Brook Centre is pictured.*

the current from the turbine and feed it onto the grid. It is part of a larger, more complicated, system (Figure 10-12) which has a metered connection to the grid. An alternative system, pictured in Figure 10-13, by contrast, is fairly simple and accepts only a fairly uniform charge from a turbine and solar panels. It does not interact with the grid, and the system as shown, has only a small inverter to power minimal workshop implements (and no power tools). This is the demonstration system that the wind and solar systems installed later will be wired into (Figure 10-14).

## Inverter options

Many sources of renewable energy are produced as DC electricity, and will need an inverter to transform the current to AC to be useful to those of us accustomed to being "on-the grid." Your inverter can tie you into the grid, and it can give you some independence from it, combined with a few batteries, if you choose the right one. If you are fortunate enough to have a utility that lets you sell back onto the grid, you could start to pay for these systems from some electricity savings.

Smaller inverters (see Figure 10-15) are useful for small loads but, for a larger system (Figure 10-16), choosing an inverter can be a complicated process. If you are off-grid, it will depend largely on the types of appliances that will be used at the same time.

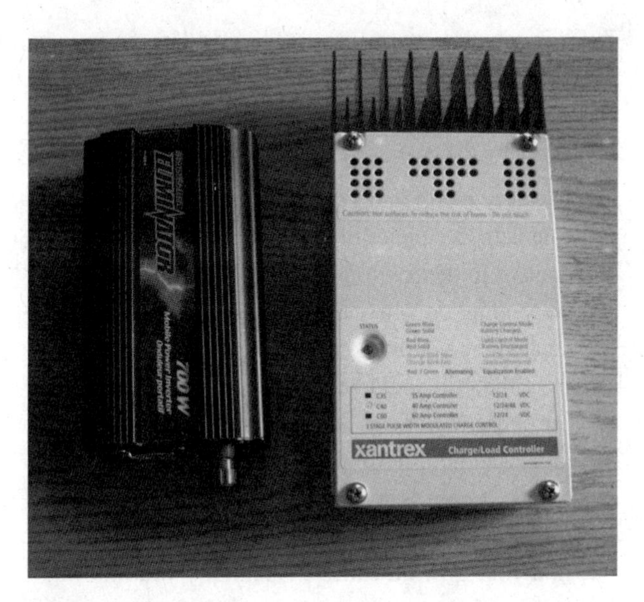


**Figure 10-15** *Smaller inverters, like those pictured, can only run a few appliances at a time.*

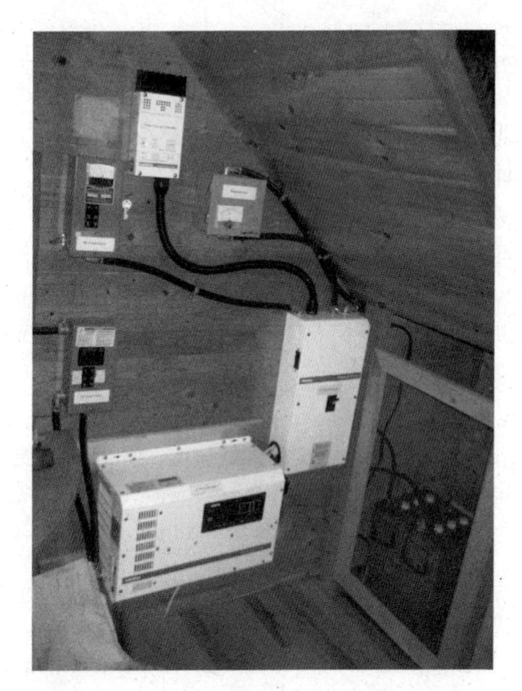

**Figure 10-16** *A larger battery-based off-grid system: batteries are on the floor on the right; the small white box to the left is the solar charge controller; further left, the big box is the inverter.*

## Project 39: Take One Room Off-Grid

Energy conservation is going to be the single most cost-effective way of reducing the costs of any battery system and reducing electricity costs and environmental impacts in general. Choosing to use the television for one hour less per day, to turn off the lights, and unplug your monitor will allow a smaller battery and lower cost. These habits are also valuable for those people who are not off the grid, and are not considering going off the grid, ever, because they teach conservation. Another way to do this would be through a meter, as we did earlier. I remember my father telling me about his days in London during the 1960s when the electricity was controlled by a meter in the hallway that he had to put coins into. For each coin he got so much electricity. Indeed, prepaid electricity cards are still common in many parts of the UK, forcing people to think about their electricity use in a very deliberate way.

### You will need

- An energy meter or to know the electricity consumption of the appliances you are planning to use in your off-grid room.
- A pad and paper.
- The electricity consumption of every appliance in your house.
- The catalog of a good deep-cycle battery provider.

## Steps

1. Add up rated (or real) consumption figures for the appliances you expect to run.
2. Multiply that by the time you would like to be able to run them. For example: taking my den off-grid would require one 23 W bulb for 8 hours, and 8 hours for 65 W laptop power: 88 W × 8 hours = 704 Wh.
3. Estimate how many days of energy independence you would like. Imagine the batteries are charged by the sun. If it doesn't shine for a week, how many days could you live without the Internet?
4. Multiply daily power consumption by days of independence.
5. Calculate battery requirements. Remember that batteries only like to be half discharged. Decide on your voltage and divide it by your energy requirements for your amp hour needs. For the example, one day of independence with a 12 V system would require a 60 amp hour battery if it was discharged all the way, really a combined 120 amp hours would be needed.
6. Calculate costs of batteries to meet current electricity consumption. (Also think about life-cycle impacts of the batteries.)
7. You can also use this number to calculate the amount of money that you could save from conservation if you are on the grid by multiplying the kilowatt hours saved by the cost of power where you live.
8. If you're really into saving the planet you can even calculate the tonnes of greenhouse gases not emitted to the environment by your conservation. Depending on where the power comes from in the area where you live, the emission intensity of power from the grid could be anywhere from close to zero for hydro or wind to as much as one tonne per megawatt hour if your power comes from old coal plants. If your power comes from natural gas it's probably somewhere around halfway between, so to find your avoided emissions multiply your kilowatt hours saved by the intensity you think corresponds to the kind of emissions for your grid.

### Tip

Don't forget 1,000 kilowatt hours (kWh) makes a megawatt hour (MWh).

# Chapter 11

# Solar Electricity

We have already discussed how much solar energy is available to us and it is plenty. Turning some of that energy into electricity is a fairly old idea but recently we have learned some new ways to do so as well.

## Electric options

There are two general ways to use solar power to create electricity. The first and oldest is to use the heat of the sun in a heat engine of some sort. Early experiments tried steam engines and air engines with moderate success. One of the challenges is getting solar energy concentrated enough to be useful to heat engines, which tend to prefer high-grade heat, around 400°C (752° F). It is possible to get solar energy concentrated to this extent but high-quality materials are needed for good performance. We discussed a model in the last chapter and commercial versions that are now coming onto the market, some of which may even be able to use low-grade heat to produce electricity.

The reader is probably more familiar with photovoltaic (PV) panels (see Figure 11-1), as their prices have dropped dramatically and popularity grown in recent years. Photovoltaic panels convert light directly into electricity. They have been around for a relatively short period of time; they are often found in small calculators and more recently are being used to power parking meters in cities around the world. Large retailers now stock ready-made solar modules that you can plug into the lighter socket of your car, to ensure that your battery receives a constant trickle charge. Changing one device over to be partly powered by solar electricity is a good way to get introduced to the ability of compact and durable PV panels. Using a very small panel to charge a battery that is charged every time the car is running anyways is probably not going far enough, but do start somewhere.

**Figure 11-1** *Frame-mounted solar electric modules at the Falls Brook Centre, Knowlesville, New Brunswick.*

Selecting a small PV module appropriate to charging the battery from the off-grid room exercise earlier is a better start, and one that teaches conservation. Batteries tend to require that the voltage feeding them is slightly higher than their rated voltage, so to charge a 12 V battery you will probably need a module that produces closer to 18 V; they are often sold as 12 V modules because they will charge a 12 V system. How long it will take to charge a battery depends on the current produced by your panel: e.g., a 180 W 18 V panel will produce approximately 10 amps (180 W/ 18 V = 10 amps) at peak production and will take approximately 15 hours of full sun to fill a 150 amp hour battery, assuming no losses and full sunlight.

Losses occur in a number of places, most often in the cable leading from the panel to the battery, and the sun does not shine all the time. At full sun, the surface of the earth receives about 1,000 W of irradiance per $m^2$ and, at other times, when the sun is at an angle to the surface of the earth, there is less direct solar energy on a flat plane. Though the sun may be above the horizon for 14 hours or more, because of its angle and the extra layers of atmosphere, etc., the number of full hours of equivalent sun is less. A tracking device can improve this by pointing the panel directly at the sun, but these devices can get complicated and expensive. To get the 15 hours of direct sunlight to charge the batteries without tracking the sun might actually take more than a full day, perhaps a few. A solution is to double the number of panels, but costs can rise quickly. Parts of the world, those near the equator, will have more hours of full sun per day than places further away from the equator, over the course of the year. The number of equivalent hours of full sun a place receives is a useful value you can look up: e.g., in most parts of North America it is common to get 4 hours or less of equivalent full sun per day over the course of a year. This means that the sun throughout the day, from morning till evening, is about equivalent to 4 hours of full sun.

To get a better understanding of how PV modules perform in different situations, there are a large number of kits on the market to help you experiment with wiring modules before pursuing a larger project. Prepackaged kits (Figure 11-2) are available, that can be assembled to, for example, power a fan (Figure 11.3) or a buzzer (Figure 11.4) as a demonstration before getting into larger projects. Small kits that might appear at first glance to be children's toys can contain enough useful gear to do valuable tasks, depending on your location, such as charge AA batteries (with an included battery holder!) to power a radio.

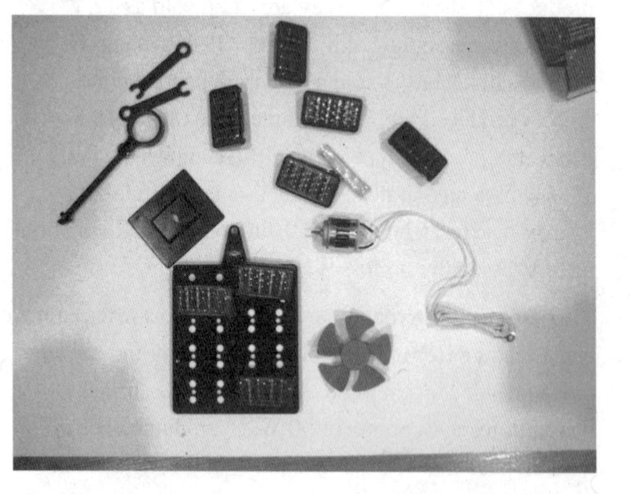

**Figure 11-2** *Content of a solar electric kit available through the Centre for Alternative Technology (see Evilgeniusonline.com to order) or probably your local hobby store.*

**Figure 11-3** *Modules wired up in a demonstration with voltage and amperage meters and powering a fan.*

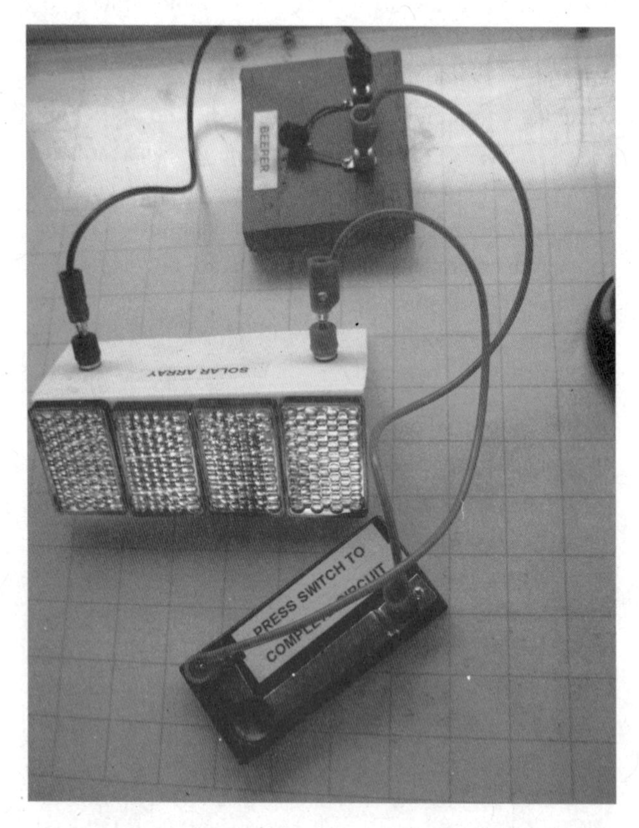


**Figure 11-4** *A solar-powered noise-maker aboard the Climate Change Bus at the Falls Brook Centre, which younger visitors particularly enjoy.*

# Project 40: Assembling a Solar Module from PV Cells

There are several places where you can pick up slightly damaged or imperfect photovoltaic cells (Figure 11-5). Surplus goods retailers on the Internet are a great inexpensive source, and some are listed in the Suppliers index. Assembling your own module has the advantage of potentially being much cheaper than preassembled solar modules, plus it is possible to integrate the cells into existing structures, like the shell of a building or a vehicle, as we'll see later, and to adjust the wiring of the cells to produce the optimum voltage and current for your application. Because you are likely to be buying slightly damaged or imperfect cells, the maximum amount of power produced by the completed module could vary from what was expected.

You will need to decide on a final voltage and current for your project, which will determine whether you'll need to cut the cells, and how you will wire them together, in series or parallel. This is a bit more of an involved project, which many people might skip in order to opt for a pre-built module. The process of configuring and installing a pre-built module can be tricky enough, depending on your location, but equally if you are in a place where money isn't very flush or where getting supplies in and out is difficult, then this might be for you! If you're not quite ready, you can follow along with one of the kits shown earlier (Figure 11-2) and assemble a few cells that can recharge a battery, for example.

**Figure 11-5** *Two stacks of cut imperfect cells ordered in bulk at a discount.*

## You will need

- Some imperfect or damaged solar cells.
- Depending on the size of module you decide to construct, you may need to cut the cells so that they fit your needs. In this case you will need a small handheld circular saw with a hard blade.
- A clean workstation with a solid desk, and good light.
- A piece of wood or other material to serve as the base. We show an ordinary piece of plywood being used to lay the cells out on; you can be creative, but be sure that you can attach a glass cover and frame to it when you are done.
- A cut piece of glass or other protective sturdy material that will fit your base.
- A frame: ours is made from reused aluminum with the corners pre-cut.
- Some silicon and hardener.
- A soldering iron and some solder.
- A multimeter.

## Steps—cutting the cells

1. The first step is to plan out your module so that you know what voltage and amperage you'll need to try to achieve. The cells we are working with are 0.5 V each and we are powering a 12 V pump for a small solar thermal panel.
2. Solar cells are rather like glass in their texture and don't take to being cut very well, so some care must be taken at this stage. Use a straight edge (Figure 11.6) that is not precious.

Figure 11-6 *Using a straight edge to score a slightly damaged cell—practice first on one that is quite damaged, to get your skills up.*

3. Rather than cutting all the way through, score the cell in a straight line where you would like it to split (Figure 11.7). The cells we obtained had two sets of contacts, so splitting them down the middle enabled us to double the number of cells.
4. Once the cell is scored (Figure 11-8), grasp it firmly (Figure 11-9) and gently snap it in two (Figure 11-10), not unlike the action of splitting a graham wafer down the center. Use your thumbs on either side of the score, and be gentle.

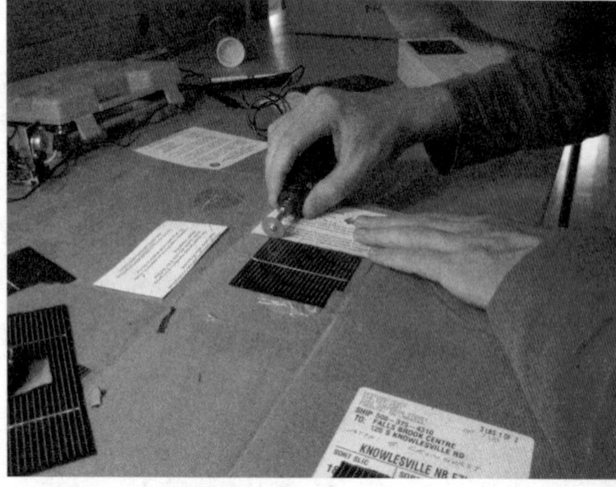

Figure 11-7 *Scoring a less imperfect photovoltaic (PV) cell in half. Be gentle, they are very fragile!*

Figure 11-8 *A scored PV cell, ready to be split.*

5. If the cell doesn't seem to want to split nice and cleanly, try scoring it a little deeper, but be careful because too deep a score can crack the cell. An imperfect split is not the end of the world; but you

Figure 11-9 *Grasp the cell firmly with both hands, and press with all four fingers on the underside and your thumbs on top.*

**Figure 11-10** *Gently "crack" the cells apart along the score line.*

should try to have a few spares in case this does happen.

6. Make sure you don't lean on the cell while cutting it (Figure 11-11) or you may crack the cell.

**Figure 11-11** *Just a little too much pressure when scoring can easily crack the cell.*

## Steps—wiring the cells together

1. Once you have cut a sufficient number of cells, lay them out face up on your plank of wood or other material, with a piece of soft protective material for now, to make sure they will all fit with a small gap (1 inch or so) around the edge of each cell. Be careful, as they are fragile and could break a little more.
2. Once you are sure they will fit, flip them over and place the electrical connections out. Most cells you will find will have a length of electrical ribbon attached to either the top or bottom layer of the cell, and a connection point to the other side of the cell (Figure 11-12). If the ribbon is not long enough, you will need to add some wire and solder a good connection to it.
3. Once you have laid out all the electrical connections and made sure that the wiring will meet your design specifications, it is time to start soldering (Figure 11-13).
4. Be careful when soldering the contacts as it is easy to put a little too much pressure on a neighboring cell and crack it inadvertently.
5. Typically, the cells will be soldered into rows, which will then all be wired together (Figure 11-14), in series or parallel as your application demands. Test each row of cells once you have finished soldering them together by taking them out into the

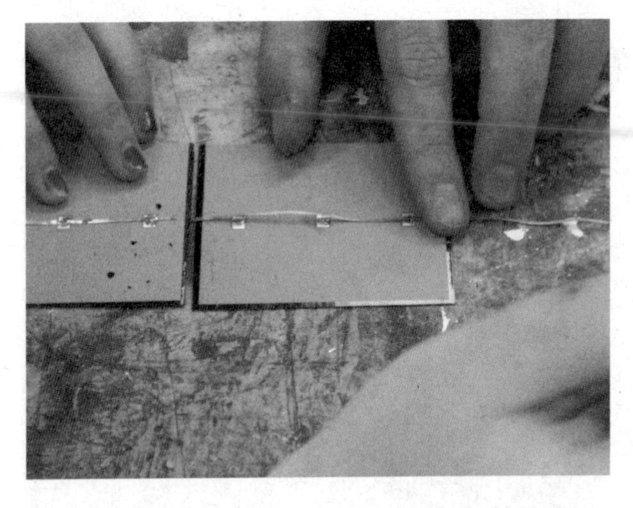

**Figure 11-12** *Lay the electrical connection from one cell over the contacts from the next.*

Figure 11-13 *Solder the connection together and be careful not to put any pressure on the cells.*

sun and measuring the current and voltage produced (Figure 11-15). Five of our strings of cells produced acceptable current, but the sixth would not, until we replaced one of the cells.

## Steps—assembling the module

1. Lay the strips of wired-up solar cells accessible, but out of the way.
2. Cover the plank of wood with an impermeable layer (a cut black garbage bag works; see Figure 11-16).
3. Place a thin layer of paper on top of the plastic (Figure 11-17).

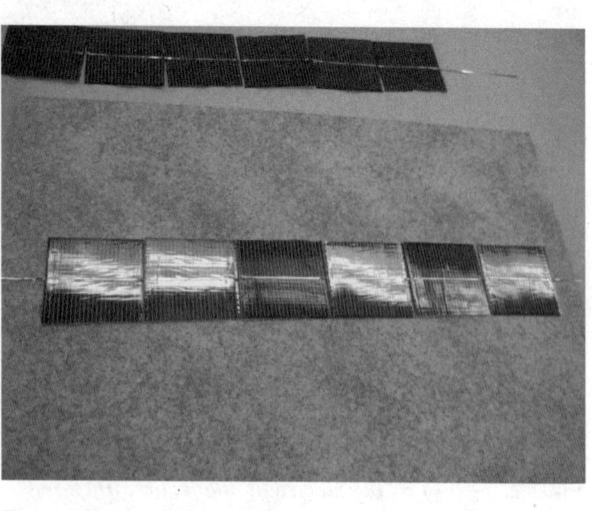

Figure 11-14 *A row of six cells soldered together.*

Figure 11-15 *Testing the electrical output of a strip of cells.*

Figure 11-16 *Prepare the base by covering it with an impermeable layer.*

Figure 11-17 *Lay down a piece of paper to protect the connections behind the cells.*

**Figure 11-18** *Solder the connections between rows of cells.*

4. Carefully move the rows of wired cells onto the base.
5. Wire the rows of cells together (Figure 11-18), being careful not to put any pressure on the cells (Figure 11-19).
6. Mix the hardener and the silicon solution in a pail that you won't need again.

**Figure 11-19** *Take care not to put any pressure on the cells and check the electrical connection before going on to the next step.*

**Figure 11-20** *Pouring silicon onto the wired module.*

7. Slowly pour the silicon out over the cells (Figure 11-20); then with gloved hands, spread it out (see Figure 11-21) evenly over the whole module.
8. Gently position the glass in place (Figure 11-22).
9. Press the glass down gently to spread the silicon around evenly (Figure 11-23) and remove any air bubbles, which will interfere with direct sunlight reaching the cells.
10. The nearly assembled module (Figure 11-24) can be weighted down (another good use for textbooks) and left to dry for 1 or 2 days.
11. When it is dry, you can size (Figure 11-25) and fit it into your frame, and test out the electrical output (Figure 11-26).

**Figure 11-21** *Spread the silicon out evenly over all the cells.*

**Figure 11-22** *Gently position the glass over the silicon and cells.*

**Figure 11-24** *Nearly finished! Let it dry for a few days under some weight (but not too much).*

## Installations

Now that you've built your own solar PV panels, it's only natural you want to think about a more permanent installation and their use. While we have been pretty casual so far about "point it towards the sun", if you are thinking about putting up a permanent, or even semi-permanent solar installation then positioning and location can have a large impact on how well your creation performs. You'll want to make sure that there is nothing shading your location during the day, and remember that shadows move, so just because a tree isn't shading your roof in the morning doesn't mean the same will be true in the afternoon. Remember also that the sun's path through the sky changes during the year, so some things might shade your collector at the time of year when you need it the most. You generally want to face your collector towards the equator and with an angle that reflects how far you are away from the equator. Many systems will have two points at which the angle of the solar collectors can be set, which are changed twice a year, before and after summer.

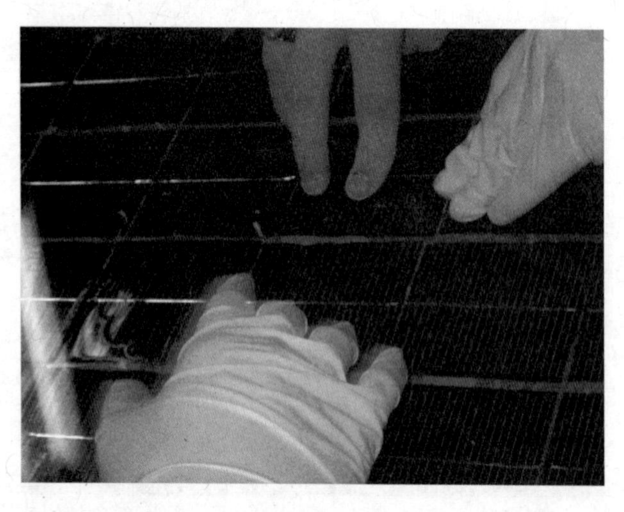

**Figure 11-23** *Push the glass firmly down to clear air bubbles from the gooey silicon.*

**Figure 11-25** *Sizing pieces of the (reclaimed) aluminum frame.*

**Figure 11-26** *The completed module in action. Powering a pump for the solar thermal collector built earlier.*

Another option is to try to track the sun, as it does move in predictable patterns, but this can get somewhat involved for a small project.

If you are going to try to add your solar system to your existing AC electrical system, you will almost certainly need to seek out professional assistance. Depending on your skills and the willingness of your contractor though, it might be possible to assist in the process to some extent. As in the case of a solar thermal system, it is the wiring and interconnection that necessarily needs to be done by a licensed electrician, while actually mounting the brackets and system on your roof is less technically demanding.

## Up or down

The choice of where to mount your PV panels is likely to be one of trade-offs. If you choose to mount the panels on the roof of your house, it could cost a fair amount to secure them, depending on the type of building. Something with a flat concrete roof will be fairly simple to fasten securely to and not very dangerous to work on, while a steep tiled roof would be more difficult and possibly require work to maintain the integrity of the roof while securing the panels and mounting brackets. Whether or not to install brackets to adjust the angle of your panels is another question of trade-offs you are likely to face. Mounting brackets can orient the panels at a better angle to catch the sun's energy through the seasons, but they can also get damaged, and may not have quite the visual appeal of flat-mounted PV panels.

It is important to choose a location where the panel is usually facing the equator if your concern is generating electricity using your PV panels. Considering the price and effort that has gone into them, I expect this be a primary consideration, but I continue to hear tales of north-mounted panels in northern countries. Too often the motivation is to look green rather than actually be green, and that is a shame because a lot of energy can go into producing a PV panel and it is a waste if that panel doesn't get the chance to produce all the useful energy that it could. Another factor of whether that energy is usefully consumed once it is produced from a suitable location, is where you plan to send it. Roof-mounted systems are nice because they are collecting energy where it is typically used—in the building. But there is sometimes only so much roof space, and a ground-mounted system is another alternative (Figure 11-27). Something to consider in this case is the length of cable run to the location where you will be using the electricity produced, as we discussed, and your electrician will tell you again, because cable losses can be significant, and vary depending on the voltage sent through them. So if you are thinking of using cheaper cable or running it a long distance, you may want to think about a higher-voltage system.

**Figure 11-27** *A wood-frame-mounted, grid-tied photovoltaic system.*

Much like the solar thermal installation, if you are going to go through the expense of purchasing PV panels to install on your house, it is likely that both the panels and your roof warrant the services of a professional. A professional installer will worry about the optimal angle to place the panels, roof loading, wind gusts, and other important things. If you have a less valuable roof, you may think about attaching the panels yourself. Do a lot of reading and use common sense before doing this yourself—maybe take a course if one is offered nearby, like the participants in the pictures that follow who are attending a class at the Falls Brook Centre in New Brunswick, Canada.

## You will need

- To have thought through your system carefully. We are only covering one part of a complete installation, and these are brief steps. You will need to think about where the electricity you are going to generate will go: either into a battery or fed to the grid if possible, or used right away. You will have wanted to size your modules to your needs and assess your options before making a sizable investment.
  - For example, the modules installed in the accompanying pictures are being fed into an existing charge controller, which came with the wind turbine installation from an earlier project. Wiring requirements are small, as the panels are mounted on the roof of the workshop where batteries are storing the electricity to power some lights and tools.
- Appropriate gauge wire to connect your modules into a panel together, as well as sufficient lengths to send the electricity somewhere useful, and various electrical connectors.
- Nuts, bolts, and mounting brackets appropriate for the roof that's taking your installation. The example pictured was planned as an educational experience using a roof that had few valuable features.

## Steps—preparing the modules

- Start by testing individual modules (Figure 11-28) so there are no surprises when they are wired together. The class pictured is installing six modules as part of a weekend Energy Experience workshop at the Falls Brook Centre in 2008.
- Lay your modules out, so the connectors on the back are accessible (Figure 11-29).
- Electrical connections on your modules are likely to be different, but usually straightforward to figure out (Figure 11-30). Make sure you have thought your system through or you will have to take it apart (Figure 11-31) and redo it.
- The frame to hold the modules together in a panel can be simple. In the case pictured (Figure 11-32) two pieces of timber per module were more than sufficient.

**Figure 11-28** *Testing electrical current from an individual module: participants at the Energy Experience Class at the Falls Brook Centre.*

**Figure 11-29** *Three modules set alongside one another; the first cable wiring them together is in place.*

**Figure 11-30** *Tightening the weatherproof covers onto the second connection.*

**Figure 11-31** *Securing wires on a second set of three modules to be installed.*

## Steps—Onsite

1. Ensure you have a safe way up to the roof, and make sure any loose articles of clothing are secure before clambering about (Figure 11-33).
2. Gather the required mounting brackets onto the roof before lifting your modules (Figure 11-34), and make sure the area is prepared. Go slow.
3. Carefully lift the modules we assembled earlier up onto the roof (Figure 11-35).
4. Lay them face down (Figure 11-36).

**Figure 11-32** *The wooden frames holding the three modules together are being installed.*

**Figure 11-33** *Try to remember to tie your shoelaces before getting up on the roof, and certainly before moving around.*

**Figure 11-34** *Gather everything you'll need.*

**Figure 11-35** *Carefully pass the panels up.*

**Figure 11-36** *Lay out the panels so that the electrical connections are accessible.*

5. Finish any final wiring, such as wiring the two panels together (Figure 11-37).
6. Carefully turn the panels right side up, making sure the electrical connections are accessible (Figure 11-38).
7. Test the panels at this point (Figure 11-39) to make sure there are no loose connections as the electrical connections will be difficult to get at later.
8. Feed the cables from your panels into a weatherproof junction box (Figure 11-40).
9. Secure the junction box to the roof (Figure 11-41) and ground it.

**Figure 11-37** *Connect the two modules together.*

Figure 11-38 *Easy for wires to get into inconvenient places, so make sure the important ones are accessible.*

Figure 11-39 *Make sure nothing came loose in the move.*

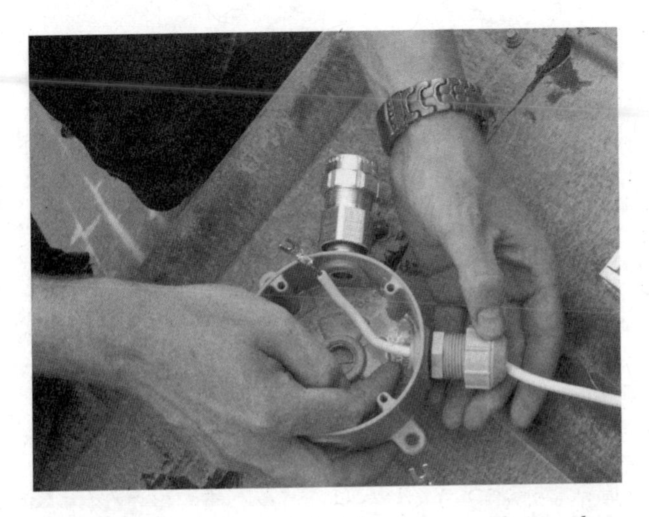

Figure 11-40 *Feed the wires from the panels into the junction box.*

Figure 11-41 *Securing the junction box.*

10. Splice the cable that will lead to the charge controller (Figure 11-42).
11. Feed the cable into the junction box (Figure 11-43) and tighten the weatherproof shield into place (Figure 11-44).
12. With the wiring basics in place, secure the panels to the roof (Figure 11-45).
13. Be careful, as your may be drilling holes very close to the panels (Figure 11-46).
14. Check to make sure your panels are secure (Figure 11-47); if not, add more support (Figure 11-48).

Figure 11-42 *Strip the end of the cable that will carry the electricity to the battery or charge controller.*

**Figure 11-43** *Feed the thick cable into the junction box.*

**Figure 11-44** *Secure the connector so the water can't get in.*

**Figure 11-45** *Time to make sure the panels don't move.*

**Figure 11-46** *Drilling this close to the panel isn't really advisable, but is sometimes necessary. Be careful.*

**Figure 11-47** *On flat roof installation, your panels are not going to have that many forces acting on them, but they shouldn't move around.*

**Figure 11-48** *Fasten every corner into a solid part of the roof, so the wind doesn't pick them up.*

**Figure 11-49** *The wires are live: make sure the ground wire (middle) is attached.*

**Figure 11-51** *Rain is bad for electronics. Fasten the junction box cover tightly.*

15. Be careful working on the wiring at this point as the panels will be producing electricity if it is a sunny day. Make sure the ground wire is in place (Figure 11-49).
16. Connect and tighten the positive cable first (Figure 11-50), then the negative terminal.
17. Close and seal the junction box (Figure 11-51).
18. Run the cable to the charge controller (Figure 11-52).
19. We should be finished on the roof: come down carefully (Figure 11-53). Remember that doesn't

**Figure 11-52** *The cable should be fed out of the way, but on the shortest path possible to avoid losses.*

**Figure 11-50** *A secure connection will mean you do not have to get up on the roof again.*

**Figure 11-53** *Climbing down after a job well done.*

quite finish the job. You still have to send the electricity somewhere useful. For this project it is connected into the charge controller from an earlier project.

20. Enjoy the view of the newly mounted panels (Figure 11-54).

## Tie or no tie

As we discussed in an earlier chapter, deciding how and if to interact with the grid is a complicated decision that gets made for all sorts of reasons. From an environmental perspective, we want as much solar power feeding onto the grid as possible. Particularly because solar radiation tends to coincide with peak electricity demand, electricity companies are starting to offer various programs that enable individuals to feed their excess power back onto the grid. If this is an option to you in your current location with your current electricity provider, then this is by far the most environmentally friendly option available to you.

Figure 11-54 *The mounted PV modules, producing clean electricity!*

The incentives could work for you too. In Ontario, for example, doing this on a small scale would enable you to pay for the average (2008 prices) solar panel in approximately 25 years, which is a very short payback period for a renewable energy system. For comparison, a similar system in New Brunswick, where production is only counted against consumption, would increase the payback period to 100 years for the average consumer. Tying into the grid ensures that none of the valuable solar electricity you generated will be wasted. If you are not near the grid, or want to ensure some level of independence, or have another motivation, then you can choose to go with a battery system. You will then need to look into inverters if you plan to use AC current, and size your battery and your loads appropriately so that as little solar electricity goes to waste as possible.

Whether tied to the grid or not, there are often times when the sun is not shining. Thankfully, we have other forms of solar energy to put to use at those times.

# Chapter 12

# Wind Power

Wind power is a valuable and in some places consistent source of energy, which we have generations of experience using. The wind is really a form of solar energy: the sun heats up some patches of air, causing circulation currents in the air, which blow across our fields and valleys. The principles behind harnessing some of that energy can be fairly simple: a cloth hung on a mast catches the wind as a sail, and fabric on a rope does the same as a kite. Early windmills had pieces of cloth hung on wooden panels, tilted at a slight angle to harness that caught wind as a turning motion useful for a variety of tasks, such as grinding corn or pumping water.

Some of those "old" ways of using the wind are still around today and some are even getting a new life because of their environmental benefits (Figure 12-1). Large freight ships are reportedly testing large kites that deploy from the front of the ship and pull it along, for use on the open ocean to reduce fuel consumption. The kites fly high in the air, because winds are more constant at higher altitudes, and don't replace the ship's engines, just improve the overall efficiency. They are not exactly like the sailing ships of old, but they are a step in the right direction. Efficiency gains are important, as is remembering that the vast majority of our global trade once happened because of the wind. When we speak of an economy powered by renewable energy sources, it is something we hope for the future, as well as something we had in the past.

**Figure 12-1** *This children's kite demonstrates an idea that people think has currency on a large scale. The turbine in the center of the kite spins in the wind, possibly generating electricity in a larger version.*

Depending on your needs and the available resource at your site, there are a number of important factors to consider when deciding if wind is the most appropriate energy source for your task. If it is, there are several options available to you, depending on what needs doing. In many cases installations will be close to electricity lines and, as we've discussed already, whether it is possible to feed electricity back onto the grid depends largely on your local utility, so is not always practical. But electricity isn't the only way to do useful work. If you decide to go down the biofuel route to fuelling your car, discussed later in the book, you might think about going so far as to produce your own oil from seeds. As in the past, grinding seeds is a fantastic job for the wind, so long as your design is appropriate. Like pumping water, grinding works well with a high-torque, low-speed sort of motion. Electricity tends to prefer a high-speed low-torque sort of motion, so the types of machines that do each job can look quite different.

Medium and large two- and three-spar wind turbines that produce electricity are beginning to dot the landscape and are becoming familiar to people everywhere (see Figure 12-2). These blades use the same effect as an airplane wing to produce lift from the passing wind and are very efficient at moderate to high winds. The shape of an airplane wing (Figure 12-3) causes a pressure to build up on one side and air to move faster around the long end of the wing, to create lift. The lift that enables airplanes to fly is directed in a circle on a turbine blade to cause it to spin.

**Figure 12-2** *A twin-blade electricity-producing turbine.*

**Figure 12-4** *A multiblade turbine.*

Windmills and water pumps tend to have many small blades, which can turn in smaller winds (see Figure 12-4), though this sort of turbine can also produce electricity. The blades on a multiblade turbine are not necessarily shaped like an airplane wing, and instead use a dragging force to spin. The blades can be fairly simple; traditionally they were cloth and wood, and are each angled slightly to catch the wind. These many multibladed machines turn at low speeds and tend to be robust enough not to get damaged by higher wind speeds. These are often hooked directly on to a pump or grinding wheel, and simply pump water to a convenient place (like an above ground store tank from a well) or grind a ready pile of seeds. It is also possible to use slower-turning turbines to generate electricity, but less common so you'd have to look into sizing your alternator properly, or some gearing. There are a number of good places to continue reading, some listed on Evilgeniusonline.com.

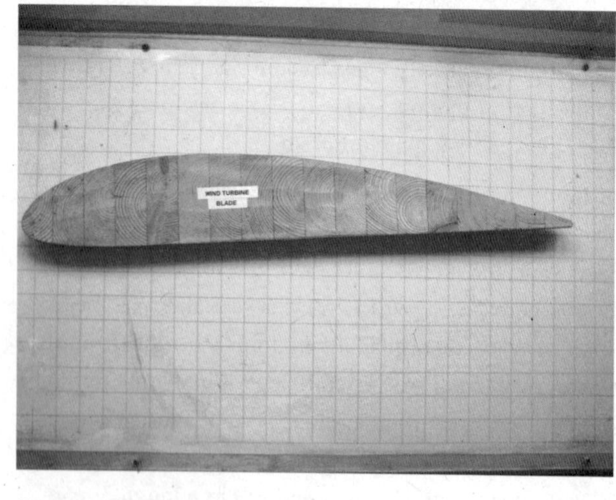

**Figure 12-3** *The same principles contribute to both airplane wing and wind turbine designs.*

All of the wind machines we've talked about have blades around an axis that is horizontal to the ground, but there is also another possibility, which has probably been in use even longer. The vertical axis turbine spins, as you might guess, around a vertical axis. Recent versions use curved blades to generate lift using the same principles as flight, whereas older versions used a drag force to capture the wind. Some of the problems that people have encountered are that vertical axis turbines are best suited for use on the ground, where the wind is not exactly at its peak. But that could be different in urban areas, where vertical lines are more common and create wind gusts, which might unobtrusively power part of our future. (Think about skyscrapers and the long upward lines along their corners.) They are already widely used as anemometers that reliably report the wind speed.

# Project 42: A Plastic Cup Vertical Axis Drag Turbine

This project demonstrates, in a simple, old and valuable way, how to capture the power of the wind. Drag is used rather than lift, to spin the machine around a vertical axis. Devices like this are used to measure wind speed today because they turn at a reliable speed in a variety of conditions.

## You will need

- Four plastic cups.
- Two plastic straws or thin wooden dowels, like popsicle sticks.
- A small wooden dowel to serve as the mast with a base, or a cork in a wine bottle, or something similar to imitate a tower.
- A sewing pin or thumbtack.
- Some tape.
- A base, or a fifth cup and a surface to tape it to.

## Steps

1. Poke holes two-thirds of the way up each cup.
2. Stick one end of the straw through the hole in one cup, and the other end in another cup.
3. Tape the straw securely in place. Do the same to the other straw and cups.
4. Align the cups so that the open end is facing the closed end of the cup in front of it (they will catch the wind).
5. Fix the two straws together at their center point with tape.
6. Push the pin through the center of the two straws, and into the top of your wooden dowel. Don't push all the way in; the cups and straws should be able to rotate.
7. Push the other end of the dowel into the bottom of the last cup (or fix it in a base).
8. That's it. Place it in the wind. The cups should catch the wind and spin it around.

## Variations

- A sturdy version of this project could be used to determine wind speed in your location before deciding on a bigger project. You can also find pre-built models in your hardware store, often as demonstrations that you could use with a small electric generator attached to the center of the turbine to record how much electricity was produced.
- You can buy commercial anemometers, or for a slightly different version of a vertical axis turbine that will produce electricity, try www.re-energy.ca/t-i_windbuild-1.shtml

# Sites and heights

Not all sites are equal where wind is concerned. Even a small change in location can have a significant effect on the overall amount of energy harnessed from the wind. The wind can be a highly variable resource. It is ideal if you are able to gather the actual wind speed data for a site for at least a year, before deciding to build a large wind-powered installation. That isn't always practical and, depending on the size of your proposed system, not always necessary. There are a number of important observations a person can make when visiting a prospective site, which can tell a great deal about the amount of wind and general direction. Generally, elevated locations will tend to have greater average wind speeds, so a mountain peak or ridge would be preferable to a valley. Alternatively, a larger tower could greatly improve average wind speeds at the turbine, but a large mast brings it's own complications.

## Assessing a site

You can tell a few things about a prospective site, just by knowing what to look for. You may know of some local areas that are particularly windy, such as a windy field or a gusty corner. For the purposes of using the wind's power, it is important to know the constancy of the wind and whether it prevails from a certain direction.

Looking at foliage or natural growth in the area can be your best indicator. Stop and look at the trees. Do the branches on one side have fewer leaves? Is the tree leaning at all? If either is true, you may have a reasonable site with a strong prevalent wind. If there are trees that are visibly leaning to one side, there is a good chance that you have a good site. You could set up a small version of the vertical axis turbine we just built, and record the speed it turns by connecting it to a data logger and small electric engine or use short periodic video captures.

## Towers

There are generally two types of mast for horizontal axis turbines: those with and those without guy lines (long cables that support the tower). The benefit of using guy lines is that they support the tower, which can be much, much lighter as a result. Consequently, they tend to be much cheaper to install with today's materials, especially for the small and medium-sized turbine. Another alternative is of course to try mounting a turbine on a building (see Figure 12-5), or incorporating it into the design from the outset. A number of innovative projects are currently underway worldwide trying to integrate wind systems into building designs. One of the biggest problems is noise and vibration, which can be an enormous and terrible nuisance to the building's occupants but may be surmountable through electronics or dampeners.

**Figure 12-5** *A building-mounted wind turbine.*

# Project 43: Raising a Guyed Tower

The typical guyed tower consists of a thin pole onto which the turbine is mounted, supported by four to eight guy wires (see Figure 12-6). The guy wires attach to four solid anchors buried into the ground, in four directions and equidistant from the central pole.

A common method of raising a guyed tower is to use a gin pole, which attaches at a right angle to the mast at one end and to a cable at the other, which is run to a pulley on the anchor. Figure 12-7 shows a guyed tower being lowered for maintenance. The tower and

Figure 12-6 *A thin tower with eight guy wires visible.*

Figure 12-7 *Lowering a tower with a removable gin pole.*

gin pole form a right angle at the hinge line, the photograph is from the viewpoint of the person letting cable out through a pulley at the anchor, to the gin pole. The turbine will reach the ground on the sawhorse in the background, in front of the small shed. In some systems the guy wires are then attached and the gin pole removed, in others the gin pole is a permanent fixture.

## You will need

- Your wits about you. We will be working with 45-foot towers, cables, and heavy electronics.
- Very likely, an experienced installer, or a very handy person around. The concepts aren't hard to understand, but the size of a tower can mean there are very real risks to those raising it.
- You'll want to have had your anchors installed beforehand. Figure 12-8 shows the rubble surrounding the anchor, the steel connection points, and turnbuckles. Depending on your site, you would probably need to dig a large hole, then install a solid anchor into a heavy mass. This is a very important step not to scrimp on, as a loose anchor could result in your valuable turbine crashing to the ground.

*Figure 12-8 The rubble holding the anchor underground and the steel connection point with turnbuckles.*

- Most commercial turbine kits come with most of the cables and accessories you'll need; a good commercial installer will bring these if not. It is often good to have extra hands around when erecting a turbine for the first time, especially ones that have an idea of what will be happening. That's what we are hoping to make of you, through this project—certainly not professional installers. We are working with a 45-foot land-based tower made by Southwest Windpower in the images accompanying the project.
- If you insist on trying to install your own tower without professional assistance, at least get the help of someone who is handy, take a class if available, read more, and talk to every knowledgeable person you can find, first. And read the instructions that came with your mast and turbine too.

**Figure 12-9** *The hinge line (vertical tube) held in place by a U-shaped rebar. The mast (down) forms a T going off to the right. The gin pole lies, detached, horizontally across the photo, just below the second U-shaped bracket. The electrical connection wires are visible where the gin pole and tower meet in the center of the image.*

## Steps

We'll be starting from the point where anchors have been installed already. The main mast joins a second steel pole (the hinge line) running along the ground and is secured to the earth under U-shaped pieces of rebar. These are loose enough to enable the hinge line to rotate, so that the mast can be raised and lowered. You can see the U-shaped rebar holding the hinge line in place in Figure 12-9. There is space for electrical connections and four equidistant anchor points with inverted steel U attachment points.

We'll raise the tower first without the turbine mounted in order to make sure that there are no problems with our setup, before we mount our valuable turbine to the top of the tower.

1. Start by laying the tower out so that the end that will have the turbine on it is elevated from the ground. Attach the cables (Figure 12-10) to the mid and upper points (Figure 12-11) on the tower.
2. Next, attach the other end of the cable to the anchor (see Figure 12-12).
3. Attach a long piece of rope to one end of the gin pole. In Figure 12-13, the gin pole (far left) has the rope (next in) tied to one end. The tower is center in the picture; at its base you can see the hole where the gin pole will be inserted, right where the cable is coming out. You might find that the tools needed vary on your project; for this one, they included a mop handle,

**Figure 12-10** *With the tower laid out, start by attaching cables.*

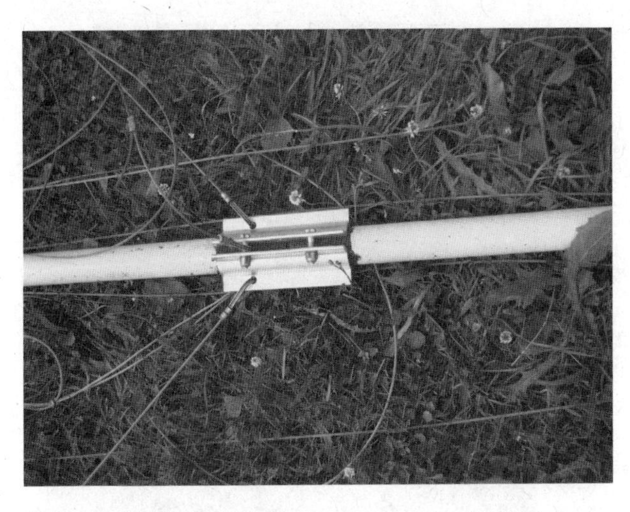

Figure 12-11 *The connection point on the tower shown in detail.*

which is used to push the wire up into the gin pole, so that it does not get pinched or broken.

4. With support ropes attached and a pulley on the lift anchor (see Figure 12-14), carefully lift the gin pole up (see Figures 12-15 and 12-16) and then gently slide it into the hole in the base (Figure 12-17), being sure not to clip the electric cable.
5. Once the gin pole is securely mounted (see Figure 12-18), we can prepare to raise the tower.

Figure 12-12 *Attach the other end of the cable to the anchor.*

Figure 12-13 *The gin pole (left), mast (center), and wire stuffing tool (right), with the hinge line and electrical connections visible at the bottom.*

6. Pulling the cord at the end of the gin pole will lift the tower (remember there is no turbine yet), as in Figure 12-19.
7. Once the tower is erect, the cables should be tight, as in Figure 12-20.
8. All went well, so we're ready to lower the mast and attach the turbine.
9. Slowly give the rope some slack to lower the mast to the ground. Have someone else guide it to a soft set down.

Figure 12-14 *The pulley attached to the lift anchor.*

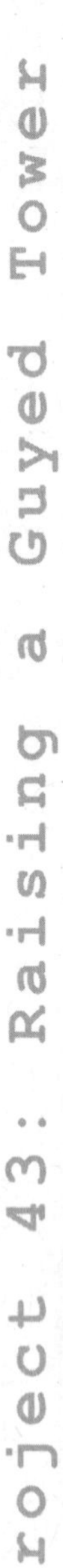

**Figure 12-15** *Raising the gin pole takes several people.*

**Figure 12-16** *The gin pole is not actually that light, pull on the long ropes to help lift one end upright.*

**Figure 12-17** *In this position, it is important to make sure the electrical cable is not pinched, which can require lifting the pole straight up.*

**Figure 12-18** *The gin pole installed.*

**Figure 12-19** *With the ropes attached, the gin pole is used to raise the tower.*

**Figure 12-20** *Safely raised, we know it works. Now we'll lower it and start installing the turbine.*

# Project 44: Installing the Turbine



Though it is not that sexy, having a mast that can be raised and lowered safely is a crucial step to having a functioning wind system. Don't be tempted to skip the last project because there was no turbine involved: without the mast this turbine is little good. The turbine pictured in this installation is commercially available, and a perfectly reasonable choice, but not necessarily endorsed more than any others by inclusion here. The turbine produces three-phase AC current, which is being used in combination with the solar module we install elsewhere, to charge a battery and power lights and fans in a workshop. The lights are DC based and the workshop is on a battery system, so the AC current from the turbine is fed into a converter, which turns it into DC current, which is fed into a battery charge controller.

## You will need

- Common sense when working around electricity and heavy electronics up thin poles.
- Screwdrivers, electrician's tape, wire strippers, etc.
- The forethought to have fed the cable up the tower; otherwise, a very long pole and some patience.

## Steps

1. Sometimes, even the best laid plans can go sour. Even if you are careful with cables, you may have to ensure, again, that they are out of the way. The mop handle works great (Figure 12-21) to push the cable up into the gin pole, and up the tower base.
2. Use tape to attach the cable to rope that you can use to pull it up through the tower (see Figure 12-22).
3. Our turbine is three-phase AC, so there are three wires coming out of the tower and three connectors, pictured in Figure 12-23.
4. Slide the wires in and tighten with a screwdriver (Figure 12-24).

**Figure 12-21** *Using a special tool to gently push the electric wire up the tower.*

5. Once the wires are attached, slide the cover closed (Figure 12-25).
6. Lift and slide the turbine onto the tower and tighten the bolts around the connection (Figure 12-26).
7. Now we're ready to lift the turbine into the air! The process should be about the same as in the last project, but heavier to lift.
8. Once the turbine is up, the lift cable will need to be moved from the gin pole to the lift anchor (Figure 12-27).

**Figure 12-22** *Attaching the wire to a guide rope.*

Figure 12-23 *The turbine electrical connection and wire coming from the tower.*

Figure 12-24 *Connecting the turbine to the wire feeding down the tower.*

Figure 12-25 *The turbine wired up, with the protective cover closed.*

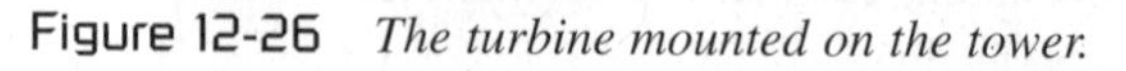

Figure 12-26 *The turbine mounted on the tower.*

Figure 12-27 *Moving the cable from the gin pole to the lift anchor.*

The combination of wind and solar electricity is acknowledged by many people with years of first-hand experience to be a fair pairing: when the sun is not shining the wind tends to be blowing. Together, they can afford a degree of independence or of constant renewable energy production to the green landowner. As mature and market-ready technologies, they are available and have a growing network of qualified installers around the world. Especially when combined with systems that can store energy for periods of time until it is needed—as we saw with chemical, water, and air batteries and as will be possible in the future with other means such as fuel cells—wind is a smart choice.

# Chapter 13

# Alternatives to Your Car

Chapter 1 described how our current $CO_2$ emissions come from different sources, like heating our homes or creating electricity. Most estimates put our transportation pollution estimates somewhere around half, if not more, of our total $CO_2$ emissions. More detailed studies show that more than half of all individual car trips are for trips that are less than 10 miles long. For those short trips the car is very often not the most efficient choice; it is after all designed for safe travel at much higher speeds. In addition, internal combustion engines tend to have a dirtier burn when they are cold, so the combined emissions from short trips are disproportionately large.

**Figure 13-1** *The Sunrider velomobile made in the Netherlands.*

In some countries where bicycles are more common than automobiles, people tend to use non-automotive means of getting around, particularly for short trips, and have far fewer greenhouse gas emissions as a result. The bicycle is among the most efficient vehicles we know of, and is ideal for short trips. In the Netherlands, Denmark, and parts of China and Africa (though with less infrastructural support), as well as hundreds of other places, bicycles are commonly used for both long and short trips. When North Americans are asked about why they don't use their bicycles more frequently, the most common answer frequently has to do with the weather.

## Getting out of the rain

There have been a great many thoughts on weather protection for cyclists over the years. If the reader was brave enough to "try another way" in an earlier project, they may have had their own already. Lately, several industrious individuals have been commercially producing human-powered and human-electric vehicles that provide weather protection (see Figures 13-1 and 13-2). Budding hobbyists have also collected a plethora of ideas on how home-builders can construct their own weather-protected human-powered vehicle or velomobile. There are also, of course, purely electric vehicles that are available, but from the point of view of sustainability, including the social aspects alongside environmental and economic factors, adding human power to the mixture has some important consequences,

**Figure 13-2** *The Leitra (right) and Quest (#1) made in Denmark.*

such as fitness and health. Just think of all the miles driving to the gym that could be avoided if all our electric-based vehicles had pedals in them!

> ### Tip
>
> There are a large number of models and designs of human-electric weather-protected vehicles available, even though they are not common on the roads. Spend some time researching before committing to building your own or purchasing one; there is something for every budget, desire, and ability.

The next few projects examine several different styles and models of velomobile—a name that is gaining popularity to describe weather-protected pedal-powered vehicles—which you can order prebuilt, build yourself, or assemble from the factory with plans and some parts. There are two general types of vehicle: those where the body provides strength to hold the vehicle together—a unibody—and those where the body is attached, either permanently or temporarily, to an underlying frame.

Recumbent tricycles are the most common vehicles to fit fairings onto, and there are a large variety of models to choose from if going down this route, as some commercial manufacturers did, as well as ways of making a tricycle that fits your needs perfectly. There are also an enormous variety of tricycles on the market, many of which can be fit with a fairing (Figure 13-3), and something to suit most tastes, including folding models, and inexpensive off-road models. *Velovision* is a popular magazine for those wanting more information. Greenspeed and ICE are just two of many popular recumbent trike makers. There are also four-wheeled vehicles, like the Rhoades Car or MicroCar by Lightfoot Cycles (Figure 13-4). Some models can fit up to four people, and lend themselves to car-like shapes when covered, as the geniuses at bicycleforest.com did (Figure 13-5).

A fairing fits around an existing vehicle, providing improved wind resistance and weather protection, but is not structurally integral to the vehicle and can often be removed. The Leitra—a commercial velomobile—uses a fiberglass fairing attached to a custom-built frame for

**Figure 13-3** *The Leitra front fairing fit to an inexpensive recumbent tricycle, named "California Sunshine." Image courtesy of Leitra.dk.*

**Figure 13-4** *The MicroCar, by Lightfoot Cycles, a four-wheeled vehicle that the company plans to produce a fairing for. Photo courtesy of Rod Miner.*

**Figure 13-5** *A four-wheeled two-passenger Rhoades Car with a lightweight fairing designed to mimic a pickup truck. Image courtesy of bicycleforest.com.*

the complete vehicle, and more recently began fitting their fairing to other commercial trikes. Another design, the VeloKit is a fabric and aluminum rod fairing that attaches to any number of commercial tricycles.

Unibody constructions, from aluminum fiberglass and advanced composites like carbon–kevlar, have a stiff body and no frame, so the wheels and seat are all attached directly to the frame. These can be very light vehicles, as Fietser's carbon–kevlar WAW (Figure 13-6) shows (a mere 23 kg (50 lb) vehicle that sings at around 50 km per hour (2 miles per hour) without the rider breaking a sweat), as well as heavier sturdier vehicles, such as the Aerorider (Figure 13-7) which encases the driver (including their head) in a bullet-like shape for electric-assisted cruising. Many people looking at the cars of the future have thought about lightweight unibody construction techniques as the way forward. Vehicles to meet most of our transportation needs are already being produced and are available as commercial products from Alleweder in aluminum (Figure 13.8), Aerorider and others in fiberglass, and Fietser and others in advanced composites. There is also an open-source velomobile project that aims to provide drawings and a forum for those interested in building their own vehicle, and countless small projects happening in garages everywhere.

**Figure 13-7** *Aerorider's traditional commuter vehicle in the background (white) and the Sunrider in black.*

## The range of vehicles

There are several different models and manufacturers of velomobiles in Europe and the variety of vehicles produced is very large. It ranges from the Twike, a $20,000 two-person highway-speed electric/pedal vehicle, to the $3,500 Alleweder self-assembly kit, right through to the free do-it-yourself book from the British Human Power Club (BHPC). Some estimate that thousands have been built by hobbyists, but the total number of production velomobiles around the world is in the hundreds.

The largest centers for velomobile use today are Germany and the Netherlands, so you're likely to be ordering parts from there or at least getting inspiration from some of the vehicles made there. There is at least one manufacturer in Germany (Cabbike) and there are

**Figure 13-6** *Fietser's WAW velomobile, made in Belgium. Photo courtesy of Fietser.be.*

**Figure 13-8** *The Alleweder velomobile, available to build yourself from a kit, or preassembled, from several sources. Photo courtesy of Dutch Speed Bicycles.*

several next door in the Netherlands (Aerorider, velomobiel.nl's Mango, Flevobike's Versatile). Nearby, Belgium's Fietser produces the WAW and the Leitra has been made in Denmark for the past 25 years. There are less than one thousand commercially produced velomobiles in the world and most of them are found in northern Europe, but a growing number are appearing in Australia, New Zealand, and North America. There are always new entrants to the market too, such as two new manufacturers in North America (Lightfoot cycles producing the Stormy Weather and Cambie Cycles producing Steve's velomobile in Vancouver) as well as a dealership in Canada (bluevelo.com) and Pedal Yourself Healthy manufacturing their improved Alleweder. Surely this list is already out of date and missing a few. The Online Resources will help you fill in my omissions.

## Some commercial velomobile links

- www.fietser.be/en/index.htm
- www.lightfootcycles.com/velomobile.htm
- www.twike.com/
- www.velomobileusa.com/
- www.aerorider.com/
- leitra.dk/
- www.cab-bike.de/
- www.flevobike.nl/content/view/25/55/lang,en/

A number of people reportedly use velomobiles for their daily commute, but there are not a lot of good data about how much a weather-protected bike changes people's travel habits. Some manufacturers direct their product and advertising towards work journeys, which allows the introduction of electric assist and accessories, whereas others are directed at the many people who also use their velomobile for long-distance trips and racing. A velomobile is not going to be perfect for every journey, but it does provide an option for a large number of common car trips. Before you decide whether this is for you, think about how many short trips you make by car. Maybe look back at the list of trips you usually make (see Chapter 2) and think about how many of those trips are under 5 km. These are the most polluting trips in your car and the ones it is easiest to use a velomobile for.

## Online Resources

- www.velomobiling.net—the North American resource includes links to all the major manufacturers around the world.
- www.bhpc.org.uk—the British Human Power Club will send you a free book *So You Want to Build an HPV* and links to the *Human Power* magazine archive, an amazing resource.
- www.recumbents.com—a large resource on a variety of human power construction projects, including velomobiles.

Though this range and availability should encourage anyone who feels the desire to be riding around in their own velomobile, there are curiously few on the road, considering it's an 80-year-old idea.

The easiest route for someone who just wants to start riding their eco-friendly vehicle around is to purchase a commercial vehicle, and there are plenty to choose from. Often the most difficult part of the decision is getting the chance to try the vehicle, at least once, before deciding to spend an average of $10,000 on it. Considering the fuel costs you will make up that price very quickly, but it is still quite a bit all at once. Many people choosing from North America will go to Europe and try several models out; the Netherlands is particularly good for trying out several vehicles in a short space of time, but that often still leaves the problem of bringing your new vehicle back to your neighborhood. Some manufacturers like Aerorider are looking into larger shipments to the United States to save on costs, once there are sufficient orders. In these cases you will have little to do but unwrap your new package and go for the first ride, at which time you'll probably feel pretty genius like and a little smug (especially if it is raining).

One of the oldest velomobile manufacturers came up with a unique solution to the issue of transportation, which was to have the velomobile fly back as luggage with its new owner. The true evil genius may of course be tempted to calculate how many short car trips made by the velomobile it might take to compensate for the flight. Consider, however, that calculating the value of new converts to weather-protected human-powered travel because of their first-hand exposure is a little more tricky.

# Project 45: Traveling with a Leitra



The Leitra has been made by hand for the past 25 years from a small factory outside Copenhagen in Denmark. Upon arrival in Copenhagen, the budding green genius is likely to be greeted by the smiling and fit face of Carl Georg Rasmussen. Spending some time at the factory once your vehicle is ready is a good way to make sure that every part of this unique vehicle fits you. Small angles on the roll bar that supports the fairing while riding may be fitted while you are there, so that it fits tight to the rider but with sufficient room to move about and pedal. If you are the least bit handy, you may even be able to assist in making parts of your vehicle, with Carl Georg's expert advice and watchful eye (Figure 13-9). For the uninitiated, this is a priceless introduction to building with fiberglass.

As with most projects so far, this is only an example of the type of challenge you may face on your journey to green, clean transportation. Other manufacturers have come up with other ways to get around the quite difficult issue of shipping within a reasonable cost. Lightfoot cycles, in Montana, has worked with the designer of the Stormy Weather velomobile Reg Rodaro (Figure 13-10) on a fairing that fits within the dimensions required for a low-cost UPS shipment. Fietser's WAW disassembles into three small pieces, while the volekit arrives in a small box. These creative solutions aim to save the customer a trip and/or only add a small amount to the final cost. Normal air-shipping charges from Europe or even across the continent could otherwise add a third of the price of an individual velomobile, but even still you could recoup those costs quickly with gas savings if getting out of the rain encourages you to use human power for at least part of your journey.

**Figure 13-9** *Carl Georg Rasmussen, builder of the Leitra velomobile, removing the fiberglass wheel cover from the mold at the factory in Denmark.*

**Figure 13-10** *Reg Rodaro with the Stormy Weather velomobile he designed, now being produced by Lightfoot Cycles in Montana. Photo courtesy of Rod Miner.*

## You will need

- The resources for a Leitra velomobile (and have decided this is the velomobile for you).
- Some large pieces of bubble wrap or other packing material (Figure 13-11).

Figure 13-11 *A well-used Leitra and the packing material (foreground) that will be used to protect it for shipping once disassembled.*

- Large pieces of cardboard or other rigid material (recommended).
- Two small wrenches.
- A plane ticket back home (this packing technique works equally well for train journeys across Canada and probably other places too).
- A few hours before your plane (or train) leaves.

## Steps

1. First, lift the front fairing of the Leitra (Figure 13-12) upwards so that you can have a seat inside, but leave the fairing up.
2. Disassembly of the Leitra starts by removing the front fairing: first removing the lightweight swing-limiter (Figure 13-13); then placing a foot under the clip (Figure 13-14) and lifting the top off (Figure 13-15).
3. To be very cautious, it is possible to remove the roof and windshield from the fairing by unscrewing several small bolts around the outside. The roof and windshield can then be wrapped separately.
4. Wrap the outside of the fiberglass in protective wrap and place it open-side up on top of the cardboard pieces.
5. Remove the rear luggage box by unscrewing two bolts (Figure 13-16) and then the wing nut in the bottom of the box (Figure 13-17).

Figure 13-12 *Lift the front fairing up and let it swing all the way forward to begin.*

6. Lift the luggage box off (Figure 13-18) and detach the bottom carriers from the frame by first unhooking them at the rear (Figure 13-19), and then lifting one side at a time off (Figure 13-20).

Figure 13-13 *Detaching the lightweight device used to keep the front fairing from rotating too far forward.*

Figure 13-14 *Use your foot while seated to unclip the front fairing from its secure hooks.*

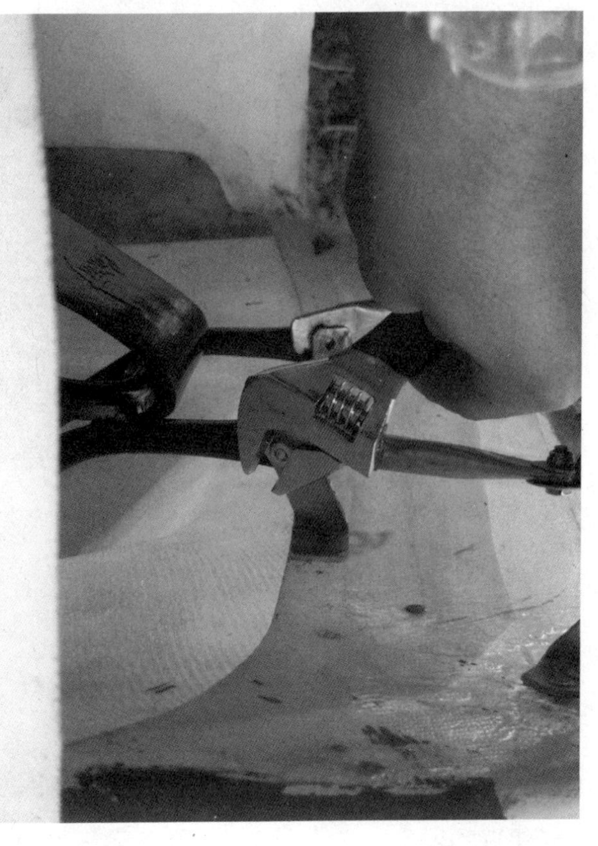

Figure 13-16 *Loosening the bolt holding the luggage carrier in place.*

7. The disassembled pieces (Figure 13-21) should get wrapped in bubble wrap. To save space, place the bottom carriers, luggage box, and lid into the upturned fairing body.
8. Wrap the whole fairing in one last protective layer; then pull layers of cardboard around.
9. Lower the roll bar on the trike and secure it with some tape. Let some air out of the tires.

Figure 13-15 *Lift the front fairing off with two hands on the aluminum support bar, and carefully place it on the protective wrap.*

Figure 13-17 *Remove the wing nut in the base of the luggage box.*

**Figure 13-18** *Carefully carry the luggage box away to be wrapped.*

10. That's it. You should now be able to proceed carefully up to the check-in counter, where you may have to pay extra fees for a bicycle. The fairing will be a large package, but because it is not heavy, should not incur any further charges.

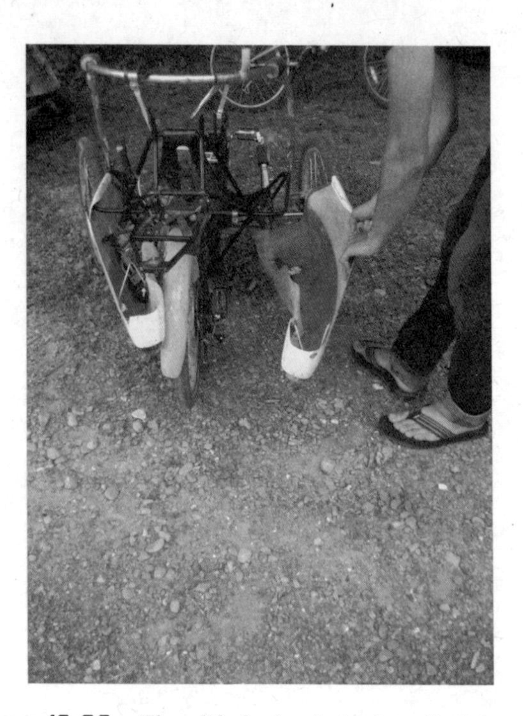

**Figure 13-20** *Then lift the hooked edge off of the frame.*

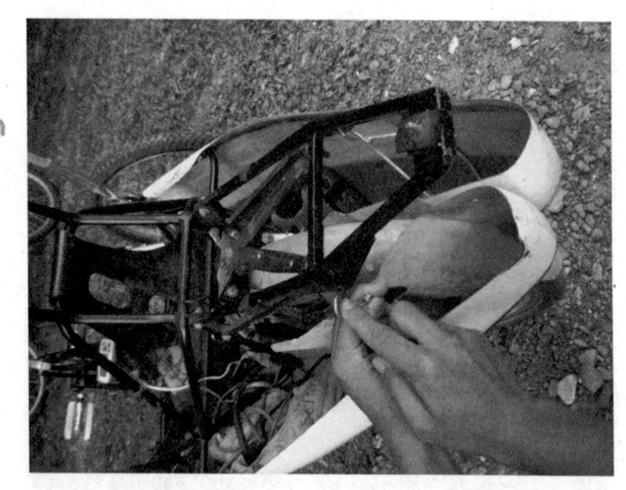

**Figure 13-19** *Unhook the lower luggage carriers at the back first.*

**Figure 13-21** *The disassembled parts of the Leitra fairing with the bare trike in the background.*

# Project 46: Frame and Fabric Tops

Probably the simplest form of some weather protection is to use a lightweight fabric and lightweight frame. Many people will be familiar with this sort of construction technique in the form of a tent for camping. Applying this idea to a vehicle is not that difficult, but does require some ingenuity. One of the most ingenious idea I have seen along these lines is the veloKit cover. Pictured in Figure 13-22, it provides lightweight weather protection and is designed to be attached to a variety of different commercial trikes. Builder and inventor Krash is keen to be able to fit the kit to a variety of trikes (see www.krash.us/velokit-trike-list.html for the list), and will offer a discount if you can supply your trike to be fitted, for those models where this hasn't already been done.

**Figure 13-22** *The veloKit fairing system, made to suit a variety of vehicles. Photo courtesy of Krash.*

## You will need

- A recumbent trike that you like and which is supported by veloKit already (or patience to send Krash your trike).
- To order the veloKit, which will arrive more or less as pictured in Figure 13-23.
- A wrench or two and an afternoon.

## Steps

1. Onto the bottom bracket (where the pedals are) you will need to attach the supplied mounting bracket. It is generally a T shape, as shown in Figure 13-23, but will vary slightly depending on your trike. When entering and exiting the vehicle, the front windshield tilts forward to allow easy access (Figure 13-24) and this is the point it pivots about, so be sure it is secure.
2. Two lightweight vertical pieces will then need to be attached to the head tube mounts. These hold the walls and side windows in place, stretching forward from the rear frame, which provides you with an accessible trunk space (Figure 13-25).
3. The trunk section attaches above the rear wheel and is made of lightweight aluminum rods that are bent into the correct shape.
4. The fabric body, expertly presewn by Krash with flexible windows and a specially blown front bubble windshield for maximum visibility (Figure 13-26), can then be attached. The kit

**Figure 13-23** *The veloKit, as it will arrive at your door.*

Figure 13-24 *How the veloKit opens to enable the driver to enter and exit.*

Figure 13-26 *The view from the inside, through the specially shaped windshield.*



leaves the bottom free so you can still put your feet down and push when necessary, but still keeps you totally dry, even riding through puddles, as any splash hits the fabric wall and not the rider.

5. To enter and exit the vehicle, simply tilt the front section away and stand straight up. Groceries and other luggage is accessible through the rear.

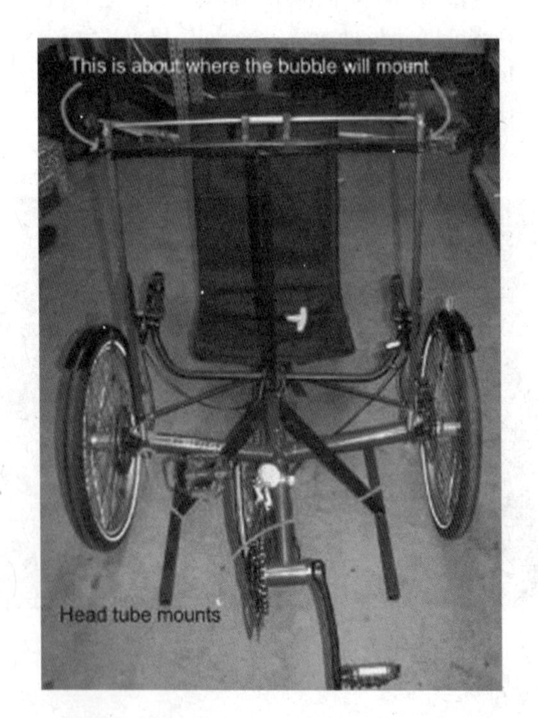


Figure 13-25 *Details of where the veloKit will attach to the average tricycle.*

## Assembling an Alleweder

The Alleweder is one of the older models of commercial velomobiles which now has several derivatives of the original design in production. Since its inception in the 1970s, it has always been available as a kit or a completed vehicle, so you can choose how much you want to get your hands dirty. There are now several different derivatives made by different companies, which share fundamental technologies and have equally fundamental differences. Most kits are designed to take about 40 hours to complete and require very few specialized tools or skills, but lots of patience and some skilled workmanship (Figure 13-27).

Figure 13-27 *Internal parts of the Alleweder velomobile under construction. Photo courtesy of Dutch Speed Bicycles.*

Figure 13-28 *Parts of the shell and dashboard of the Alleweder under assembly. Photo courtesy of Dutch Speed Bicycles.*

The Alleweder is formed by aluminum riveted together in a sort of bullet-type shape (Figure 13-28). There are several variations on the original design now available from manufacturers in Europe and the United States: check the Online Resources to locate some. Do some reading from people who already built them before deciding on one. Regardless of which version of the Alleweder you decide on, no welding is required in the assembly of the kit. The difficult pieces have all been precut and it is designed for someone with "only one left hand," but no other special skills.

### Online Resources

- www.dutchbikes.nl—sells the original Alleweder kit from the Netherlands and ships worldwide.
- www.pedalyourselfhealthy.org—manufactures and sells Alleweder derivative kits in the USA.
- www.dutchbikes.nl/uploads/builders_manual_KV4_1.6%20h.PDF—complete instructions for assembly.

## Project 47: A Zotefoam Fairing

For the home builder, the process of creating fiberglass molds for a single vehicle production may seem labor intensive, and in search of a solution one man has been successful in using Zotefoam—a closed-cell cross-linked polyethylene thermoplastic material—to easily create a stylish and practical fairing. Of course, there are countless other ways to do this; fiberboard (like old election signs) is another popular choice and the pieces are easily attached together using duct tape. Zotefoam can be a bit trickier than cardboard and duct tape, but the end result has many advantages and the process involves fewer molds than a fiberglass fairing.

### Online Resources

- www.recumbents.com/mars/pages/proj/tetz/projtetzmain.html—John Tetz's building page with links to the *Zotefoam Manual* and a massive resource for homebuilders on the rest of the recumbents.com site. Many thanks to John Tetz for images and assistance preparing this project. The reader interested in this project is urged to visit complete directions at his site before proceeding on their own project.
- www.ihpva.org/—The international human-powered vehicle association has a useful builders corner and links to their mailing lists, an invaluable source of advice for those in the process of conceiving or building their own vehicle.

## You will need

- A trike or other suitable vehicle you intend to attach a fairing to, and a creative idea for how to attach the two.
- A male mold or some rigid foam (that pink stuff—common 2 inch thick home insulation) and an idea of what your fairing should look like (the BHPC book is a good place to start).
- A heat source, preferably several—to heat-form the Zotefoam to the male mold—such as electric space heaters or a large air dryer.
- Bungee cords to attach the foam to the mold while it is heated and cooled.
- A small amount of fiberglass and resin to add strength to sensitive areas such as the nose cone and small pieces of metal to brace the fairing in shape.
- Contact cement to weld pieces of Zotefoam together.

## Steps

1. Start by making a drawing and model of your trike and how the fairing will fit over it. Remember to consider things like the distance the wheels have to move when turning, and how you will get in and out.
2. A male mold has about the same shape as the outside of your fairing will have, and its surface should be relatively smooth. Many hours can go into tracing a smooth curve on layers of insulation, gluing them together, and sanding it all flat, but it need not be exhausting. Be sure to wear protective clothing and be in a well-ventilated area when sanding (see Figure 13-29).
3. Lay out the color pattern your final product will have, and order sufficient Zotefoam for the purpose and some extra. Use masking tape as a guide before trimming the foam; then use bungee cords to attach the Zotefoam to the mold. If you are going to place fiberglass in the nose cone, now is the time to do so (see Figure 13-30).
4. Arrange your workspace so that the heaters can be suspended about 18 inches above the foam; be adjustable and able to move along the mold

Figure 13-29 *The male mold ready for sanding.*

Figure 13-30 *Fiberglass in the nose cone for a little added rigidity.*

Figure 13-31 *The top color of Zotefoam laid out and fastened to the mold.*

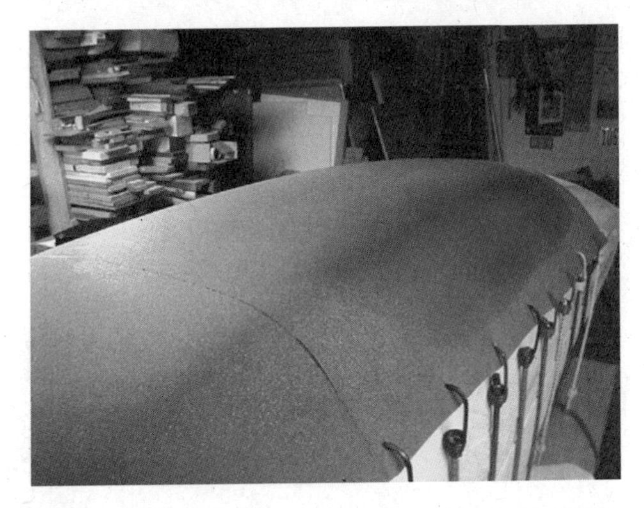

Figure 13-32 *The change in color is visible at the far end of the fairing.*

Figure 13-34 *The fairing, once removed from the mold and cooled.*

(or move the mold along the heaters), and leave your hands free (see Figure 13-31).

5. Once the colors are laid out, start by forming one color piece at a time. The heaters should be kept moving at all times while heat-forming the Zotefoam, with slow even movements. The foam will change color slightly as it is heated; noticing this can help you prevent overheating, which can spoil your material. Practice helps you notice how much heat is required to form but not burn the foam (see Figure 13-32).
6. Contact cement can be used to join two pieces of heat-molded Zotefoam together into a fairing. You'll notice that this light body is still floppy, and you will want to build some braces to help it maintain its shape (see Figures 13-33 and 13-34).
7. Depending on your model of trike, it can be tricky to mount the fairing. With some models, you have to remove the wheels from the trike to attach the fairing, but it is likely that each individual project is different (see Figures 13-35–13-37).

Figure 13-33 *The two colors joined after having been heat-formed separately.*

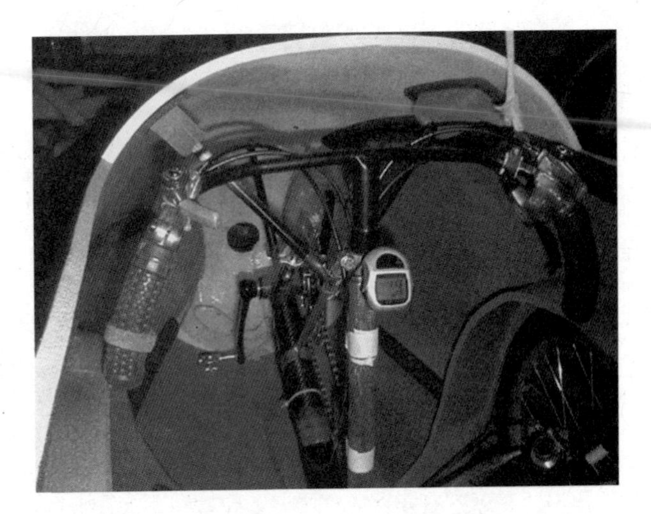

Figure 13-35 *The inside of the fairing, once it is mounted.*

**Figure 13-36** *The partly mounted Zotefoam fairing.*

**Figure 13-38** *The relaxed sitting position of recumbent vehicles makes them more attractive to actually use.*

## Even lighter

A plethora of ideas for smaller and lighter vehicles are available that can meet some of your transportation needs. For adventurous individuals looking at building their own vehicles, we've sort of glossed over frame building, because it can be very involved. Atomic Zombie's *Bicycle Builder's Bonanza* by Brad Graham and Kathy McGowan is a fantastic introduction to the subject with heaps of great ideas for innovative ideas for useful and fun vehicles, and can help you keep your car in the garage more often and through more seasons.

Part of the problem with using a car for so many of our trips is that cars really do weigh a lot. It is quite silly to be taking a heavy machine around, just to move a single lightweight person around most of the time. The velomobile starts to get at that problem by trying to provide some of the comforts of a car without the weight. But there are often times when you might not need a car or all of its features: e.g., when the weather is nice, and the sun is shining. Recumbent trikes, which have a relaxed comfortable ride help (Figure 13-38), while two-wheeled electric scooters (see Figure 13-39)

**Figure 13-37** *The fairing near the wheels—notice the small clearance.*

**Figure 13-39** *Narrow and light electric vehicles are especially appropriate for urban areas, but also face many regulatory hurdles to becoming street legal.*

**Figure 13-40** *The BionX electric assist system is a popular choice for cyclists. It replaces the rear wheel in a very simple operation that can be performed on most bikes.*

are becoming popular for people who do not find the exercise of a bicycle something they need, as are electric assist systems for regular bicycles (see Figure 13-40). For those who want something a little larger, there are smaller automobiles such as the Smart Car on the market, but the real genius is urged to fully consider such a choice. One effective way that cities have found to reduce the number of automobiles circulating is to restrict parking, making people consider one more element of inconvenience before choosing their car for a short journey. The Smart Car removes that inconvenience, and may actually make using a car for short journeys more attractive and increase individual emissions.

# Chapter 14

# Alternative Fuels for Transport

Earlier in the book, we looked first at ways to reduce your car use by using the tools already available to you. In Chapter 1 these included carpooling, using your feet for short trips, and using public transport. We also looked at some simple techniques for maximizing the fuel economy of your existing car. In Chapter 13, we looked at alternative vehicles that might make you a little more comfortable for those trips where you leave your car at home, and some lighter alternatives for other trips.

In this chapter we focus on the fuel you put into your car and explore some alternatives you might look into. These are of course longer-term projects, which people newly inducted to the world of green might not immediately be willing to do. They are perhaps something to think about for the future, recognizing that making modifications to existing technologies is a good way to reuse, but is also a stopgap. Creating an engine that can use a variety of fuels easily, as we saw with external combustion engines, is not technically challenging. Today it is often other factors, such as our dependence on diesel and gasoline engines, that can hinder technical change to the engines we use, which have a long life. In this case fuels can be tailored to the engine. However, some of these "alternative" fuels have problems of their own, which everyone should take the time to consider before deciding whether it is a route they should put the effort into trying.

## Biofuels: a question of balance

Ethanol is currently all the rage in North America. It has been promoted as an available renewable fuel that could help kick the oil habit, and increase the profitability of farms around the country in the meantime. Biodiesel has come a close second in the race to being perceived as a viable alternative fuel for our transportation needs—one that also has the potential to be used in many other locations where heavy oil is used, like in home heating. Ethanol can be made from a wide variety of biomass. In the United States it is largely made from corn, whereas in other parts of the world such as Brazil, sugarcane is used. Biodiesel can also be made from a variety of raw biomass sources, and different sorts of vegetable oils are often used.

Both of these sources of renewable fuel for our vehicles present some very serious problems, as currently produced. The reader is urged to exercise caution before deciding to go down this route. Both ethanol and biodiesel use a raw material that is frequently sold as a food product; the impact that the growth in popularity of these fuels could have on the world's food supplies is likely to be significant. This is so serious that a United Nations expert recently called biofuel use a crime against humanity (see www.monbiot.com/archives/2007/11/06/an-agricultural-crime-against-humanity/ for more information). Now, as we discussed, we need to cut our energy bills and the amount of carbon from sources deep under the earth's crust. The question is, do biofuels do this?

The biofuels energy balance is something that scientists and energy analysts are debating. Estimates try to take into account all of the inputs to growing crops, including making and applying fertilizers, as well as the energy that goes into fermenting and distilling the fuel. Many of the estimates seem to conclude that every barrel of ethanol uses a little more than a barrel of oil to make it; some claim that it is a little less. The energy balance with biodiesel can be slightly better, often because waste oil is used and because the oil does not need to be fermented. Ethanol does tend to burn slightly cleaner than gasoline and, as it is sold in the United States currently, as an additive to gasoline, it could help an engine's emission performance.

But can these new fuels help us reduce energy? Barely, if at all, as currently produced, and especially not if used in our current large and wasteful vehicles, with our current wasteful driving habits. Also, any energy reduction has also to be measured against the very real possibility that putting a food crop into our cars today may mean that others are going hungry. Rising corn prices in 2008 following an ethanol boom reflect the reality that there is only so much arable land on the planet, and if we use too much of it to grow fuel for our too-large trucks, some people could go hungry. This may even be true if one is using a waste product, such as used vegetable oil, because of how our economic system works. For example, some waste vegetable oils are currently collected from Canada and sent to parts of Asia, where they are used as a chicken feed. Now, the reader may question whether it is a wise use of resources to send used oil several thousand kilometers away, and whether we like to have our chickens eating used oil. That is fair and neither of those could be considered terribly wise. However, the result of increased use of waste oil in North America could mean that there is less cheap chicken feed in Asia. This is likely to result in increases to the price of chicken feed in Asia, with the consequence that fewer people are able to afford to eat chicken.

A deep-green environmentalist, who has recognized the link between animal consumption and energy use, might not mind fewer chickens consumed by people. Animals use more energy for each unit of energy they deliver as a food product, compared to vegetables. A low-energy diet therefore includes little or no meat in it, but it is sustainable and just that each person should be able to make that choice for her- or himself. A person having too little food to eat so that another may drive a large car with a clear conscience about the fuel they are using is not a fair solution. Indeed, it is hardly a solution.

Nonetheless, we'll show you how to make biodiesel and convert your car to straight vegetable oil (SVO). Why? Not necessarily because we need more vegetable oil cars on the road—we need fewer cars on the road more urgently—but because the process of getting close and dirty with the fuel you use is an incredible learning experience. There are also possibilities for advanced biodiesel and ethanol fuels, grown from algae or using crops that are inedible and grown on marginal land, that may make these biofuels a more socially responsible option in the not so distant future. There are also other fuels that could be used and this is just a small taste of the possibilities that I'm sure the true evil genius will explore further. During the depression in the 1930s, when fuel was difficult to afford, it is rumored that the waste product pictured in Figure 14-1 and the bulky solid fuel in Figure 14-2 were turned into methane-rich gases that were successfully fired in automotive internal combustion engines; more on these options in a future volume.

Creating one's own fuel from waste is among the most vivid educational tools I have encountered. It excels at making people aware of their energy consumption, though a similar experience may also be had cutting and collecting your own firewood. The sweat equity developed by the process, and the realization of the quantity of fuel that is typically required for the average lifestyle, are both valuable. It is far too easy to fill up the car with a hose coming from a pump, and never really realize that 20 liters (a little over 5 gallons) is too heavy for a person to carry, never-mind ponder where it came from or what it took to get it there. Going a step further and growing the crop and pressing the oil would bring an even deeper connection, but dealing with the sights and smells of waste can also be its own worthwhile experience.

**Figure 14-1** *Human and animal manure, if processed correctly, can produce a gas suitable for use as a transportation fuel.*

Figure 14-2 *Wood can be processed into a methane-rich gas, which can then be used to power spark-ignition engines for transportation purposes.*

## Vegetable oil and biodiesel

What is known today as biodiesel is the product of the transesterification of vegetable oils in order to reduce the viscosity of the oil. We need to do this because the modern-day diesel engine is not designed for vegetable oil, but for a thinner fuel we call diesel. We call it diesel because of Rudolph Diesel, an inventor who lived about 100 years ago and came up with the idea of an engine that ran on the basis of compression. The engine he demonstrated used a vegetable oil to run when he unveiled it to the world, and his idea is said to have been to enable farmers to grow and process fuel for their own needs. Not long after, a petroleum-based substitute was found and popularized, and we now call this fuel diesel, for some reason. Because this diesel fuel is thinner than your typical vegetable oil, modern engines have been optimized to use a thin fuel. It is not a difficult engineering challenge to create an engine that would run on straight vegetable oil, indeed older diesel engines already can. Until that vehicle is available though, the fuel can be thinned.

There are two well-known ways to make vegetable oil thinner, or decrease its viscosity: one is a chemical reaction that produces biodiesel; the other is by heating the oil before it reaches the engine. Through a process called transesterification vegetable oil reacts with methoxide, which we create by mixing methanol and lye, to produce biodiesel and glycerin. This reaction was first discovered by a chemical company looking for a way to produce glycerin, which is useful for making bombs and soaps, but for most biodiesel makers the glycerin is the waste product.

### Online Resources

- Journey to Forever is one of the most valuable resources on the web for the intrepid renewable fuels explorer. Their biodiesel recipes at journeytoforever.org/biodiesel_make.html have been used by hundreds of people across the world.
- A video of the process of making biodiesel with common household products is available at www.videojug.com/film/how-to-make-pop-bottle-biodiesel.

# Project 48: A Blender Batch of Biodiesel

The process of making biodiesel is not very difficult, but it does require some patience and careful attention to the instructions. It is in the intrepid evil genius' interests to do as much reading and hands-on learning on the subject as possible before attempting this alone: this is meant as a brief introduction. There are as many different ways of making biodiesel as there are colors in the rainbow, and lots of people have shared their experiences online.

## You will need

- Some new or used vegetable oil (Figure 14-3). For the first batch, start with new vegetable oil to make life a bit easier. A good small batch to start with if you are using a blender for a processor is about 1 liter in size.
- A means of bringing the temperature of the oil to 50°C (122°F). In a larger system you will want to permanently install this either in a preheating tank (Figure 14-4) and/or in the main processor (Figure 14-5). For larger batches, it is also important to be able to maintain the heat of the oil for a longer period of time. For smaller batches, this is not so important and a small gas or electric element and pot will do the trick to warm up the vegetable oil.
- To track down methanol (Figure 14-6), you will need about 20% by volume as much as the vegetable oil you are planning to use, and 99% pure. You can sometimes find this at car racetracks, as it is a fuel for some racing vehicles. You might also find it in products that clear the fuel lines of gasoline engines, but read the ingredients carefully: 20 ml will do for a 1 liter sized test batch, but do some reading on the web before trying this at home.
- Lye catalyst. Only a small amount is needed: 5.3 grams of 97% pure potassium hydroxide (KOH) or 3.5 grams of 97% pure sodium hydroxide (NaOH). Lye absorbs water quickly, so do not leave it exposed to the air for long periods of time.

**Figure 14-3** *Gathering and using waste oil can be an inconvenient as well as a dirty job.*

**Figure 14-4** *The biodiesel processor at the Falls Brook Centre in New Brunswick. Two filters above the heated black barrels are fed by a pump near the floor, which circulates and filters the oil constantly as it is heated.*

Figure 14-5 *The main biodiesel reactor has a water jacket that maintains the required temperature during the reaction of methoxide and vegetable oil. The glycerin and then biodiesel is drained from a valve at the bottom, and the tank is fed via a pump through an inlet at the top.*

- A processor. On a large scale this needs to have a heating element capable of bringing the oil to 50°C (122°F); on a small scale it can be a kitchen style blender, but an old one as you will not want to use it for food products again. In a pinch you could use a 2 liter pop bottle and shake it with your arms rather than run the blender, as you will only need to do that for a few minutes anyways.

Figure 14-6 *A container of methanol with a custom delivery spout.*

- Methoxide mixing chamber: an HDPE (high density polyethylene also known as #2 Plastic is resistant to corrosion) bottle or other vessel (Figure 14-7) that will not corrode in contact with methoxide and will not be affected by the exothermic (produces heat) reaction. The vessel should, ideally, have a tight sealing lid and a method of mixing it (on a small scale, shaking will do) as well as a means of ensuring that the toxic vapors produced by the process are kept well away from the operator. Make sure you have appropriate safety equipment, including corrosion-resistant gloves, safety goggles, and a filtered-air mask.
- Scale, accurate to 0.1 gram.
- Funnels.
- Duct tape.
- Thermometer.
- Using used oil batch would require titration equipment, which would include (Figure 14-8):
  - a pipette
  - a beaker
  - an eye dropper
  - acid test strips or phenol red
  - Distilled water and some of the lye.
- Finally, a container to let the biodiesel settle out in, if you don't want to be scraping glycerin from the inside of your blender (a 2 liter bottle will do, the same one as before if you skipped the blender altogether).

## Steps

1. There are several good recipes for biodiesel on the Internet: most vary to some degree, but still work. There are also many detailed books on the subject, and this is really just a brief introduction.
2. You will want to start by warming up the oil to about 50°C (122°F). While this is going on (Figure 14-9), move onto other things, but keep an eye on it.
3. The titration equipment is used to measure the acidity (free fatty acids) of the oil in order to determine the amount of reactant needed to make a successful batch of biodiesel:

**Figure 14-7** *The methoxide mixing vessel (on the right), which is vented and feeds directly into the main reactor pictured in Figure 14-5.*

- Measure a small sample of vegetable oil into a beaker (Figure 14-10) and add phenol red.
- Slowly add droplets of lye/water (0.1% NaOH) mixture from the pipette (Figure 14-11) and keep the beaker moving in a circular motion to stir the liquid.

**Figure 14-8** *A pipette, distilled water, and other materials required for a titration.*

**Figure 14-9** *Warming the used oil in this large-scale system includes filtering it and continuous pumping to ensure a good mix of the oil.*

- Note the volume added at the moment when the reaction is visible (Figure 14-12) and has completed (Figure 14-13). Repeat it a few times, and use the average to calculate the extra amount of lye you will need in order to ensure a complete reaction, if you are using waste vegetable oil.

4. With the oil warmed up, measure and prepare the methanol and lye (Figure 14-14).

**Figure 14-10** *Measuring oil into the beaker.*

**Figure 14-11** *The beaker placed under the pipette with a measured volume of distilled water added slowly.*

5. Slowly add lye to the methanol, and mix the methoxide carefully for at least 10 minutes.
   You will feel the container start to get warm. Ensure you don't inhale any of the fumes from this reaction as they are hazardous to your health.

**Figure 14-12** *Note the volume of reactant added when the reaction starts to take place.*

**Figure 14-13** *Stop adding reactant when you can see the color of the liquid inside the beaker change.*

6. Once the methanol and lye have formed methoxide, slowly mix it in with the preheated oil.
7. Mix the whole solution for about a half an hour before letting it settle and cool (Figure 14-15).
8. Once it settles, if the reaction was successful, you will see a dark layer of glycerin at the bottom and a lighter layer of biodiesel above it (Figure 14-16).

**Figure 14-14** *Wear protective equipment when handling lye, as it is caustic.*

**Figure 14-15** *A small amount of completed vegetable oil and methoxide mixture, before it has settled out.*

**Figure 14-16** *The process of glycerin settling out to the bottom of the jar, leaving biodiesel above, is clearly visible.*

9. Filter the biodiesel before putting it into the tank of a diesel car. Unlike the SVO conversion covered next, no conversion of your vehicle is required. Do be aware that using home-brewed biodiesel will void the warranty of any new vehicle, so use common sense. A failed reaction can easily destroy an engine too, so be careful testing your fuel before using it.

## Straight vegetable oil in your car

To save you the trouble of brewing a batch of biodiesel for every tank or two, it is possible to modify an existing vehicle to run on straight vegetable oil. By incorporating a filtration system and pump into your vehicle, it is possible to pull up to waste oil bins, which are located at the back of nearly every restaurant, and fill up your tank. Keep in mind that doing so without permission of the restaurant owner is stealing, so always ask first. Also, remember to ask what was going to happen to that oil. It is not a bad question to ask the restaurant owner, after he has said you could take it, of course.

### Online Resources

Commercial vegetable oil conversion kits include:

- www.goldenfuelsystems.com/
- www.greasecar.com/
- www.plantdrive.ca/html/vegtherm.html
- www.greaseworks.org/svo.

And other home-built examples and hints can be found at:

- www.dancingrabbit.org/energy/winter_svo.php
- www1.agric.gov.ab.ca/$department/deptdocs.nsf/all/eng4435.

# Project 49: Tearing Up Your Fuel System and Feeding in Waste

For a fairly moderate cost of the conversion kit, which can be constructed or ordered as a retail item, it is possible to drastically reduce fuel bills. What sort of kit and how many modifications you will need to make will depend on the climate you are operating in. Warm climates simply need less heat added to their vegetable oil-based fuel to make it thin enough for a diesel engine. Vegetable oil can also be mixed, in small amounts, directly with diesel fuel, and there are a number of reports of people simply pouring fresh oil into their tanks in warm climates with not too many mishaps. In cold climates, vegetable oil has a tendency to get thick and sometimes congeal, especially if it has been used before, so properly heating the oil before it reaches your engine will keep you moving.

**Figure 14-17** *Two metal jerry cans repurposed as vegetable oil tanks in the trunk of a VW Jetta. The tank on the right provides extra storage capacity and holds a prefilter, which enables raw grease to be collected from restaurants. The tank on the left is heated by copper pipes filled with radiator fluid, and has a fuel line running through the radiator fluid up to the engine.*

## You will need

- Patience and probably another mode of transport to use for the time this small surgery on your vehicle is performed.
- There are a wide variety of techniques used in the commercial kits, and an even wider variety of ways to convert your own vehicle. The concept is generally to have a second tank (Figure 14-17), so one tank can contain diesel fuel that is used to start the engine and create some heat either through the cooling system or by powering an electric heater. For some extreme climates, it is often advisable to do both (as is the case pictured, from Edmonton, Alberta).
- The average kit will have a second, heated tank, or expect you to be creative in finding one. Try to have good seals around the tank (unlike the accompanying images), as you will be heating used vegetable oil in it, and the smells of its previous use can tend to be exacerbated by heating, and seep into all manner of fabrics in the vehicle and worn by its occupants.
- The tank can be heated by electricity or by tapping into the vehicle's cooling system or exhaust system.

  There are advantages and disadvantages to each method: an electric heater might be easy to install; a radiator fluid system can very simply be copper pipes in a salvaged jerrycan; while using the exhaust saves the trouble of running pipes of hot fluid down the length of a vehicle (as in Figure 14-18).
- Insulated hoses, or for cold-weather systems a hose-in-hose system that prewarms the fuel all the way from the tank using the radiator fluid as a liquid "jacket" surrounding the fuel line, run to the engine compartment (Figure 14-19). The fuel line will need to be cut into, and a solenoid (Figure 14-20) will be inserted to enable you to switch between the two tanks. A diesel engine does not consume all of the fuel that is fed to it, and typically returns some fuel to the tank. To save on costs, it is possible to "T" the return line on the vegetable oil side back into the engine, also reducing the amount of diesel fuel that gets mixed with your vegetable oil.

**Figure 14-18** *The hoses filled with radiator fluid and, through the center in its own tube, heated vegetable oil, running through a hole inside the vehicle.*

- To try to plan out your system well before starting to cut hoses and drill into your vehicle to secure the tanks. Clamps, tools, a soldering iron, and blowtorch are common tools you'll need if you aré going the route of building your own conversion. If you're not comfortable with these sorts of tools, you'll probably want to think instead about ordering a kit and having your local mechanic install it, with your input so you understand how the driver's behavior will need to change. For example, with an SVO system, it is necessary to run the engine for a short period of time on diesel fuel before shutting the engine off, in order to ensure that there is diesel in the fuel system when it comes time to start, and not congealed vegetable matter. This requires the driver to remember to switch back to the diesel tank a block or two away from the destination, especially if the vehicle will be parked for a long time or in a cold place, otherwise starting will be difficult. There are some new products that try to handle this process in their own way, so do some research if, for example, you think this might be a problem.

**Figure 14-19** *The extra addition to the radiator system is about center in the photograph. Two T junctions tap into the existing system and feed hot fluid down to a second set of T junctions, through one of which runs a tube filled with heated vegetable oil.*

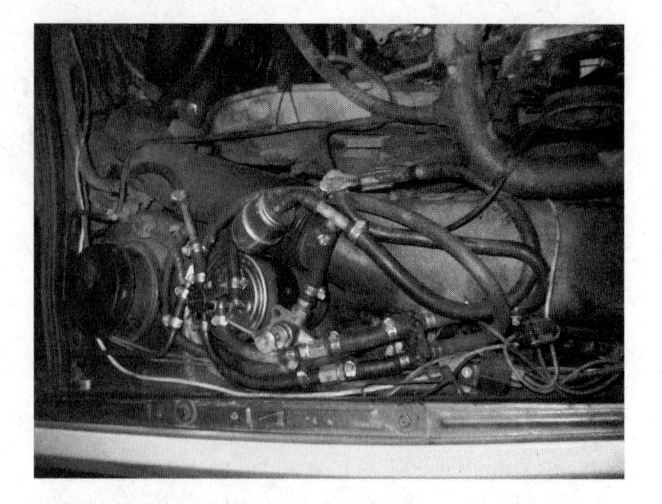

**Figure 14-20** *The location where the fuel line has been cut into, and the six-port solenoid (a little left of center) in the photograph. An emergency return valve has also been fitted to this system to enable air bubbles to be cleared after a filter is changed.*

## Steps

1. Read much more on the subject of what you intend to do and understand the very real risk that you could damage the engine of your vehicle beyond repair by converting it to run on a fuel for which it was not designed. Any manufacturer's warranty would of course be void, and your insurance status may technically change. In most countries any

vehicle on the road is required to pay a fuel tax, and, by using a substance which has not had taxes paid on it as a road fuel, it is likely that the upcoming conversion will put your vehicle in violation of tax laws, unless you take corrective actions. The UK has a tedious process in place that vegetable oil vehicle owners should use to calculate and pay their tax bill. In other countries where this is not in place the responsible genius might think of bringing the matter to the attention of his elected representative and using the opportunity to discuss the importance you place on good climate change policies with them.

2. It is often easiest to start by installing the tank, as this step does not necessarily disable the vehicle. An electrical system would have the relatively trivial task of running electric wires to a heating element in the tank, whereas the radiator fluid-based system pictured earlier has a more involved installation process that includes splicing into the vehicle's cooling system. It is advisable to ensure that the heating element in the tank and hoses leading to the engine compartment are completely free of leaks before attaching them to your vehicle's cooling system. A leak in the tank can bleed your vehicle's cooling system dry in a matter of minutes and easily damage your engine by overheating.
3. The fuel line leading from the tank should be protected from damage and, if in a cold climate, well insulated or fed through the center of the cooling fluid hose.
4. A fuel line heater (Figure 14-21) can serve as anything from a sufficient heat source for your whole system, if you live in a hot country, to a supplementary heater for already heated oil, as shown in the images using the Vegtherm electric fuel line heater. There are various types of fuel line heaters available on the market, including ones that use the radiator fluid to heat fuel in a compact device. It is a good idea to think about extra filters (Figure 14-22) on the vegetable oil-fed side of your fuel system, and many people have also come up with heated fuel filters to ensure that congealed oil does not clog the filters.
5. Once your fuel line is correctly attached to a three- or six-port solenoid, you will need to wire it up to a convenient switch, accessible from the driver's seat (Figure 14-23).

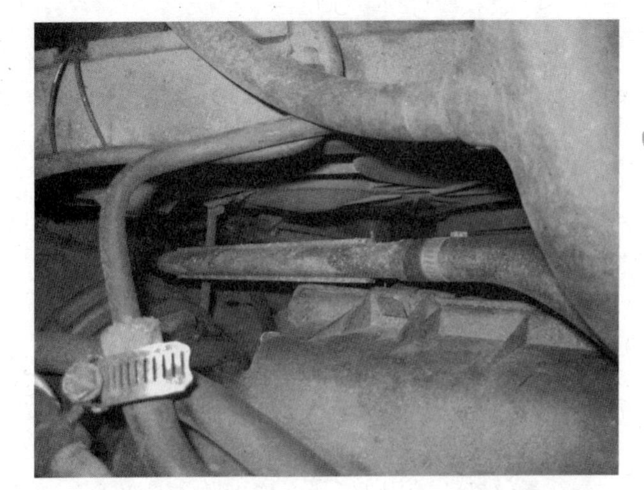

**Figure 14-21** *The Vegtherm electric fuel line heater (above center—also visible in Figure 14-20) provides a secondary heat source on this cold-weather system, but could be sufficient on its own in a warmer climate.*

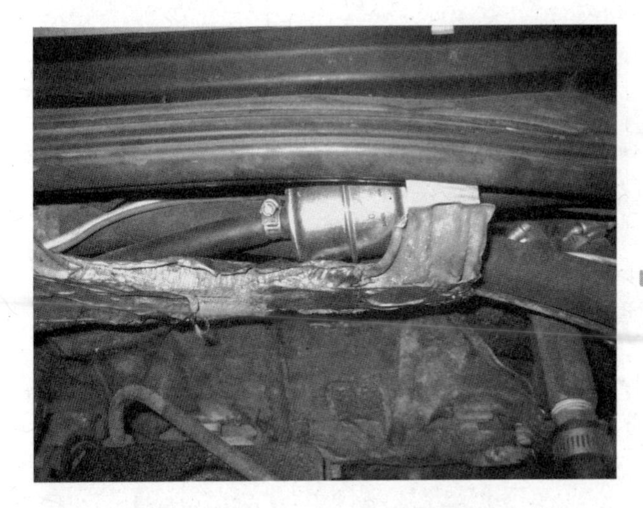

**Figure 14-22** *Second and even third filters on the vegetable oil side of your fuel system help protect your engine from small pieces of food waste.*

**Figure 14-23** *The electric switch that will connect to your solenoid and change the fuel source from diesel to straight vegetable oil. You will need to remember to turn it back to diesel a few minutes before arriving and turning the car off, or else starting again will be a problem.*

**Figure 14-24** *Of course the decals are optional, and always remember to use alternatives—like the Leitra velomobile behind the car—when practical.*

6. In a very brief series of steps, that is the basic concept. The VW Jetta pictured (Figure 14-24) was expected to take a day for the conversion operation, once the homemade system was designed, but it took about 6 weeks in total to work out the problems that inevitably arose. *From the Fryer to the Fuel Tank* by Joshua Tickle is a priceless book for the aspiring vegetable oiler, and a must read before actually proceeding on a conversion. There are also many other valuable books on the market.

# Chapter 15

# Spaces and Structures

This book began talking about transportation, because its emissions are the biggest single part of our emissions. Concentrating on the easy fixes first, we then moved on to discuss some of the harder projects you might try to green your means of transport further; e.g., by building weather-protected human- and electric-powered vehicles. We've also discussed the ability to use our cars on alternative fuels, and the dangers that some of these fuels could pose, not just to the environment but to other people too. In addition to transportation technologies, the book has also touched on greener ways to produce electricity, e.g., through hydro or wind production, as well as greener ways to heat and cool buildings.

We have tended to focus on the individual actions and choices; however, too often, the choices we as individuals are able to make are limited by the world around us and choices that have been made by the planning of others. The reality of the transportation problem we face is that infrastructure increasingly locks us into our existing patterns of travel, which tend to be very high-energy consumers. The same goes for housing; increasingly, large suburban developments lock us into a high-energy-consuming lifestyle. There are options, but the farther we get down the path of building suburbs, freeways, and multiterminal airports, the harder it is to move towards less high-energy-consuming ways of life. In this chapter we will look briefly at what citizens can do to change the way the structures and cities in which they live are built.

## Structures

If you are starting a new project, it's a good idea to think about energy early on. For a new construction project, choosing the site can have a large impact on the energy consumed in transporting materials to the site, on the renewable energy resources available onsite, and on availability of transportation links to surrounding services and communities. If you are constructing an office, for example, where people will travel to every day, it is a good idea to locate it close to an existing stock of housing or a major transportation route. Otherwise, the office could indirectly result in the construction of houses, roads, and possibly all the infrastructure that follows development, including power lines, sewage systems, highways, etc.

Before considering new buildings, it is always a good idea to consider reusing an existing structure because there are frequently environmental and social benefits to doing so. Unfortunately, the economics don't always reflect that, so like other environmental decisions, they have to be made for reasons other than money. Nonetheless, try to find a site with the best transport access and existing infrastructure for your needs, which infringes on natural areas as little as possible. Existing sites have already disturbed a natural area, and tend to have services close by—so they are a natural choice.

Onto your site is likely to come a foundation of some sort. Most times today that will involve concrete, in order to comply with building regulations (Figure 15-1). Concrete, although it has a great many uses, has a high amount of embedded energy, owing to the huge amounts of energy that goes into its production. Embedded energy is a concept that captures the energy which has gone into production of a resource, such as mining the gravel (Figure 15-2) and transporting it to the site. It is efficient, if we are using a high embedded energy material, to use it wisely. Try to minimize your use of concrete if possible, through judicious design and alternatives where appropriate, while still being aware of the benefits of such a heavyweight material: this not only refers to how much concrete weighs but also to its thermal performance. We talked about how water and stones can hold a lot of heat energy earlier; a building is the perfect place to put this to use. Try to think about the orientation of a building as if it were your solar

**Figure 15-1** *A heavy concrete foundation beneath a lightweight wooden wall construction is a common building technique. If well-insulated, this method of construction can produce buildings that are efficient.*

cooker, with a glazed southern exposure to maximize the amount of the sun's heat that can be trapped if you are in a northern area. The sun's energy can then be usefully stored overnight and sometimes longer by using a heavyweight material in the part of the building being heated by the sun. If you are in a warmer climate, heavyweight materials can be used to mitigate fluctuations in temperature by being cooled with fresh air at night, and radiating their cool through the day.

**Figure 15-2** *Gravel is a component of lots of things, including the concrete in most foundations, and is usually mined in large open pits like this. The concept of the embodied energy of the materials you use assesses a portion of the energy consumed by such an operation to the materials you use.*

## Composite structures

A composite structure is one that is made of constituents that remain recognizable. Typically in the built environment, the structural properties of the different parts combine to work as a whole. Composite structures can be found in a large variety of applications, from reinforced concrete in skyscrapers, to glass-reinforced plastic in car and boat bodies (Figure 15-3).

"Sandbag" structures are just one type of composite structure, which can be called a contained composite structure, because one of the materials (the sand) is completely contained by the bag. The bag can be filled with nearly any material, in which case its name should change; however, in practice sandbag often refers to bag-type buildings filled with a variety of materials. "Earth bag" is another term gaining in popularity, which is a slightly more accurate description of a composite structure that is filled with earth, often gathered on the building site itself. They are a priceless introduction to the building process that can be pursued on a small or large scale, without enormous cost.

## How sandbag structures are built

A sandbag structure is made by filling a sack, bag, or other container with sand, and then laying the filled bags on top of one another to form the desired shape.

**Figure 15-3** *Composite structures are all around us, in the hull of the boats and reinforced concrete in skyscrapers.*

The procedure is more or less identical for earth bag construction, where the bags are filled with earth, often from the building site itself.

All contained composite structures utilize the combined properties of each constituent material to attain their structural strength. The inner material provides the mass of the structure and presses against the containing layer, expanding its shape to create structural strength. A friction element between layers of bags is sometimes used to decrease slippage and increase the structural strength of structures.

The resulting structure has properties that neither of the two materials could have on their own. Sandbag structures have commonly been used in the past as an emergency fortification or shelter because of the strength of the structure and the ease with which it can be built, from materials that are sometimes to be found on site. As a temporary structure, the sandbag method has many advantages, which are not lost when used to build a more permanent structure.

The main deterrent to using a contained composite structure for a permanent structure is the deterioration of the containing layer, which could destroy the integrity of the structure unless accommodated for by using an appropriate material for the outer layer. Sandbag structures in the past that have used natural fibers like burlap as the containing layer have encountered this problem, which is still encountered using more modern materials like polypropylene, especially if it is exposed to sunlight. For polypropylene sacks, a commonly used adaptation is to render the outside of the building with a protective layer that will halt the deterioration of the structural strength of the building.

Another common adaptation is to use a material on the inside of the sacks that will bind to itself and solidify once the bags have disintegrated and ceased to contribute to the structural strength of the building. This method, used in the adobe style of building, eventually ceases to have the characteristics of a composite structure once the bags disintegrate.

A particularly innovative method of sandbag buildings is called the super adobe system. It uses a single long bag instead of the typical series of smaller bags, along with both measures already discussed, to prevent the deterioration of the structural integrity of the building. Super adobe buildings are often designed as a dome, which is a shape that is particularly strong when built with most contained composite materials, not just super adobe, because of the way that the forces of the roof and walls are spread evenly throughout the structure, all the way to the ground.

### Online Resources

- www.calearth.org—Nader Khalili is the founder of the California Earth Institute and pioneer super adobe designer and builder.
- greenhomebuilding.com—a skillful collection of resources, including details of how to build some beautiful insulated earthbag domes.

## Thermal properties of contained composite

A contained composite structure will get its thermal characteristics largely from the thermal characteristics of the material that is used to fill the containers. Earth and sand will of course exhibit thermally heavy characteristics, while lava rock—which has many holes in it—might be more insulating because more air is trapped. Because an earthbag-type structure uses a large number of smaller pieces to form the structure of the whole building, and to protect the bags from deterioration, a final rendering layer can also contribute to the thermal qualities of the building.

## Sandbags as heavyweight and lightweight materials

Sandbags are interesting as an introductory building material because they can perform as either a heavyweight or lightweight material. Heavyweight and lightweight structures can perform similar functions, but they do so in completely different ways, as we've already discussed. A heavyweight building moderates temperature differences by absorbing and releasing heat energy from either side, whereas a lightweight building operates by maintaining a temperature difference on either side.

Sandbags can be used to build heavyweight or lightweight buildings by varying the containing material. This can be performed by completely replacing it with a material that has a higher insulative value, by mixing two materials to increase their insulative value, or by building a layered structure.

An earthbag structure will have many of the heavyweight characteristics of other styles of construction with earth. Heavyweight buildings work by retaining thermal energy and releasing it slowly, rather than maintaining a temperature difference as an insulative, or lightweight fabric, would. Optimally, a heavyweight building will have a temperature delay of approximately 12 hours, meaning that the building will release the heat from the previous 12 hours. This is optimal, because the times of day when the heating and cooling requirements of many buildings are highest are approximately 12 hours from the times when the daily temperature cycles are at their nadir.

Heavyweight materials can be very useful from the perspective of the thermal performance of your building, particularly if they are planned for at the outset of a design. A common criticism of thermally heavy materials is that they tend to be made of concrete and other materials that require a lot of energy to produce and use. That isn't always bad, because they are strong materials that will be used for many decades to come. Like cheap fuel though, having a plentiful source of an inexpensive material can tend to encourage people to forget the alternatives. And there are some alternative heavyweight materials, like sandbag and earthbag building techniques, to experiment with right under your nose.

## Project 50: Building with Earth

Earth building has a lot of potential to be an educational experience, with several interesting building principles.

### You will need

- Some dirt and clay (Figure 15-4).
- Large planks of wood to create a form (Figure 15-5), if you would like to experiment with a rammed earth wall, as the accompanying pictures do, and potentially an arch form for an exploration of structure.

Figure 15-4 *Mixing earth on a cardboard plank.*

Figure 15-5 *A simple wooden form can be used to create a thermally massive earth wall to store heat and cold and reduce energy costs.*

- Bags. Lots of bags. Burlap or plastic will do equally well, but there are other materials on the market, some specially designed to resist corrosion, for example. You will have to adjust your building to the bag or the inside material if you expect a structure to last, perhaps by mixing cement with the earth, or by protecting the bags from the elements. Thankfully, this is not an expensive material, so it is possible to do many experiments, even testing a structure through the seasons, if you have the space. You could experiment with structures that are not bags too; as shown in Figure 15-6, relatively weak materials, like aluminum cans, can have some structural strength in a composite structure—in this case, wrapped in chicken wire.
- For a larger structure you may consider using barbed wire in between layers of bags, mixing earth with hardening agents to prevent structural failure if the bags erode, or covering the walls with a render such as lime or clay, to prevent deterioration of the bags.

Figure 15-6 *Two arches made of composite materials—on the right is an earthbag arch and on the left an arch made from reused aluminum cans and some chicken wire—shows the potential strength in individually weak materials.*

## Steps

1. To start with a small wall of sandbags, dig up some dirt (or rocks or other material), place it in the bags, and stack the bags in a shape you would like. You can easily make a square shape from a sandbag structure without a form. It is often used for flood defenses, because it has an enormous amount of strength as a composite material (both the bag and dirt contribute towards the final properties).
2. Rammed earth building compresses the earth into a form (see Figure 15-7) that can be very strong and support one or more stories. Earth can be compressed by hand or with ram tools (Figure 15-8), and then the form is removed.
3. There are many techniques for doors and windows. Making smaller rammed earth bricks with a brick maker (Figure 15-9) is one option.
4. You might also want to try building an arch. To begin, build your form from wood. It should resemble the arch you would like to build, only slightly smaller. When stacking around the arch, place your form, then stack bags evenly up both sides, with a locking bag at the very top (see Figure 15-10).

Figure 15-7 *A rammed earth wall built as an educational experience by students at the Centre for Alternative Technology, in Machynlleth, Wales.*

Figure 15-8 *Ramming earth into a wooden framework to create the earth wall shown in Figure 15-7.*

Figure 15-9 *An earth brick maker works by compressing a small amount of earth into a small brick, which can be used to create all manner of structures.*

5. Sandbag structures can be very strong, support a lot of weight (see Figure 15-11), and last a long time, particularly if covered or plastered to avoid corrosion of the bags. Another alternative is to mix cement in with the dirt so that the heavyweight structure has its own strength once the bags disintegrate. The Online Resources lists people who are using this technique and others who are experimenting with lighter-weight materials in an effort to raise the insulative properties of this material.

Figure 15-10 *A sandbag arch. Note how there is one central bag, directly in the middle, which forms an almost triangular shape. This is called the locking bag, because it locks the arch into a sturdy form.*

Figure 15-11 *A properly constructed arch, made of sandbags, can support a surprising weight. Strong, standing room structures can be built environmentally sensitively and inexpensively using this building style.*

## Project 51: The $500 House

It is easy, especially when looking around at very expensive eco-homes, to think that everything green has to be expensive. That is simply not true. Some very inventive people are not only thinking about affordable eco-housing but also putting the plans up for free, for everyone to read and see.

The hexayurt (Figure 15-12) is a fantastic experiment that can be done on both a small scale and a large scale. The deliberately open-source and public domain plans include large to small houses, intended as easy-to-build structures relevant for refugee populations. The hexayurt has a number of interesting features, including scalability, a modular nature, packaging with a range of renewable energy technologies, and the use of common materials that are quick and easy to work with. The other low-impact materials we cover in this chapter do not tend to have the last feature listed (easy to work with), though that is the predominant factor in material choice by the building industry.

Depending on how long the unit is expected to last, several different materials can be chosen for the construction. The outer materials have been chosen to be easily available, durable, easy to work with, able to moderate the indoor climate, and be esthetically pleasing.

## Online Resources

- hexayurt.com—the main source for information on this project.
- www.appropedia.org—the main wiki page for the project.

For a permanent shelter, a builder or potential manufacturer might consider the following materials:

- Permanent use: Thermax HD (Dow).
- Temporary use: laminated hexacomb cardboard (Pregis).
- On-site fabrication: Tuff R (Dow, widely used).

There are three shelter sizes; the middle size is shown in Figure 15-12. The houses are designed to be very affordable. For example, the size shown is expected to cost around $200 before any services. Full-sized units take a team of three people around 1 hour to assemble. They are assembled using a 6 inch wide, 600+ lb bidirectional filament tape, and anchored to the ground like tents. No heavy lifting, ladders, or scaffolding is required.

## You will need

- The complete free plans are available to download from www.appropedia.org.
- Sheets of material that are used on a large scale are insulated and have a layer that reflects heat, which is useful in the hot areas where the structures are expected to be most useful. The hexayurt is a fairly simple but very useful observation on shapes, which is that 4 ft × 8 ft pieces of material—which is a common size for several industrial materials—can be cut in half and assembled easily to form a shelter form, with minimal material waste.

**Figure 15-12** *The hexayurt—an open-source environmentally responsive building technology that costs little to build and can be constructed quickly.*

- It is relatively simple to follow the designs with sheets of cardboard that are twice as long as they are wide (4 ft × 8 ft).
- This material is held together with an advanced joining device, called sticky tape.
- When combined with low-impact technologies—like solar panels, battery chargers, efficient stoves, etc.—and aimed at migrant populations, such as refugees, it has the potential to provide high-volume, low-impact, and affordable serviced shelter—see appropedia.org for more details.

## Steps

1. This is actually a remarkably simple idea to put into practice. A major trouble when working with large square sheets can be that a lot of material is wasted. The designer of this project thought about this and the form reflects conservation of materials.
2. Five or six pieces form the walls of the hexagonal house. Twelve triangular pieces form the roof, and tape holds them all together (see Figure 15-13 for a simple version). Standard 4 ft × 8 ft pieces are rearranged to form a habitable structure.
3. On a large scale, the industrial material is insulated already. All that remains is to wire and plumb it in. Already covered earlier, you should be experts by now.

**How to Build a "Pup" Hexayurt**

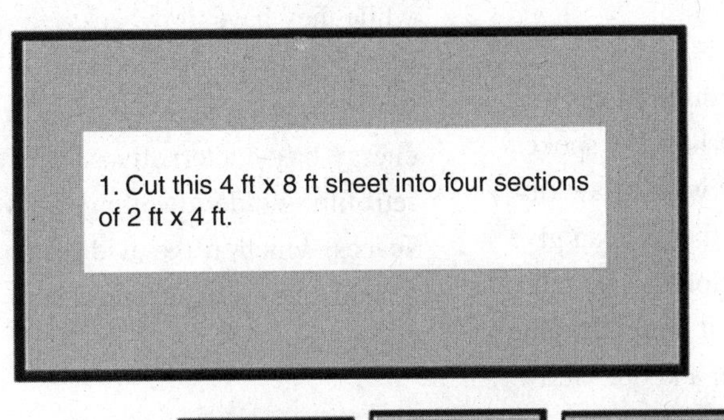


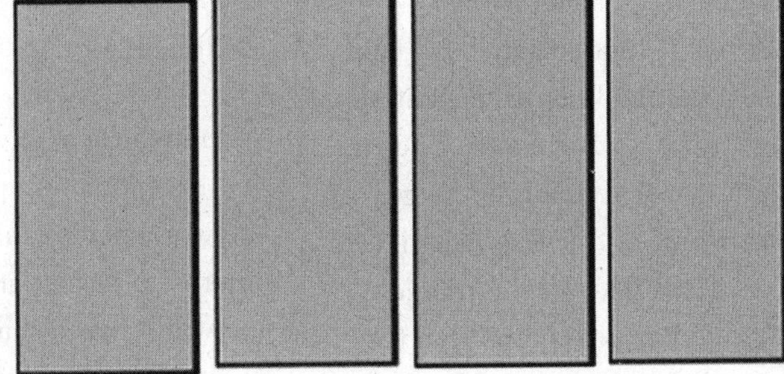

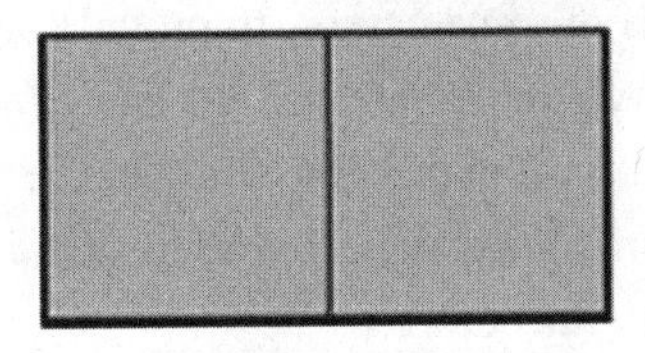

2. Cut three of the 2 ft x 4 ft boards in half. This leaves you with six 2 ft x 2 ft squares.

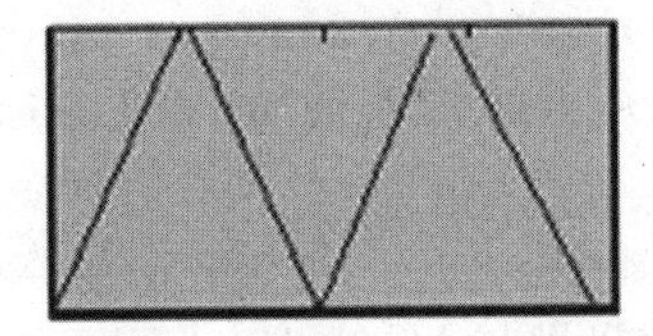

3. Cut your remaining board into the three isosceles triangles. Join the two scrap pieces so that you get a fourth whole triangle.

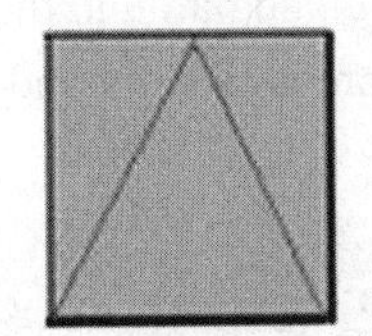

4. Take the sixth 2 ft x 2 ft square wall board, and cut it to get one whole isosceles triangle, and two halves that you will join. This gives you a total of six isosceles to join together for a hexagonal roof.

5. When you join the five remaining 2 ft x 2 ft squares and attach the roof, you'll have a five walled "pup" hexayurt which is approx. 3 ft high.

**Figure 15-13** *How to build a "pup" hexayurt. Plans for a cardboard-based model-sized pentayurt that will help the reader to understand the building ideas presented. They can be downloaded for free from the Appropedia site.*

## Building with straw

Straw comes up too often, I think, in a discussion of environmentally friendly building materials. We spoke of waste and value earlier, and in many ways straw fits the bill very well; it is a waste product that we can give renewed value to by using it for insulation. I have seen first hand how difficult it can be to use it correctly. One of the biggest problems is air leaking in and out. Straw doesn't particularly like to sit straight and, though compression by using it as a structural material can help to maintain tight barriers against the floor or foundation and roof, over time it can require regular maintenance to keep up its thermal performance. This tends not to get done. In addition, it can be a time-consuming material to build with, which makes it less than ideal for large-scale implementation. It also needs protection from the elements in the form of a good foundation and good roof, otherwise water can quickly destroy its integrity. The reader considering this route is urged to take hands-on lessons from experienced builders before embarking on a larger project. Leaky structures can have an enormous impact on resource consumption, and straw structures are no exception, despite their positive qualities as a low-impact building material.

## Other options

There are a simply enormous number of new environmentally sensitive building materials on the market, with more arriving every day. Many are easy to work with and have a relatively low embodied energy content, like sheep's wool batting, whereas other more traditional alternatives, like expanded foams, might be more easily accessible and familiar to installers. Manufactured products, including insulation and paints, can release gases and pollutants that can often be harmful to the occupants of a building long after they are delivered and installed.

Volatile organic compounds (VOCs) are released by paints and fire-retardant sprays commonly applied to building materials and home furnishings, and can provoke allergic reactions in some people. Alternatives are sometimes available as a choice of paints and some other products, but often require some searching and research. For insulation, traditional foam materials, while they have drawbacks, save an enormous amount of energy once installed, and should not be overlooked while awaiting more convenient or lower embodied energy-based alternatives. This is especially true when retrofitting older dwellings or working with limited spaces, which, if the avid reader has taken considerations about travel seriously, is more likely to be the case.

## Spaces

As true today as it was 20 years ago and—aside from a few exceptions—a pattern that the many cities around the world have followed for the last 50 years is to give priority to faster patterns of movement. Planners around the world have been bending over backwards for the automobile—narrowing sidewalks and removing bike lanes in some cases—to provide an unimpeded path for the venerable automobile. There have also been pushes for airport expansions all over the world, with more communities subjected to their noise and air pollution than ever before.

It wasn't always this way. Before planes and cars, and almost gone from memory, is a history of wide-reaching train networks that provided cheap and reliable transportation on the ground. Streetcars (Figure 15-14), as they were and are still known in Toronto, San Francisco, Istanbul, and around the world, move thousands of commuters daily and are a clean and

**Figure 15-14** *Among the first streetcars in the world, this nearly 100-year-old streetcar is still operating today in the bustling city of Istanbul, in Turkey.*

efficient way to move people about. But as the automobile became ever more popular, streetcar lines, across all of North America and other parts of the world, were torn up and public transport services in cities were provided by buses instead.

Buses are not necessarily an inconvenient way to travel. There are very comfortable overnight buses that have beds running lengthwise down the coach, but these tend not to operate in the high-energy consuming continent of North America. In Canada, Europe, and many other parts of the world, it is possible to get a bed on a long-distance train, as well as a freshly prepared meal. While this is possible in Canada, at a comparatively steep price, to my knowledge the United States no longer has such a service. In circumstances like this, flying or driving can look like the only option for long-distance travel, and that this is the case cannot surprise many.

On a local level the tendency of developers to build sprawling suburban developments, creates an over-reliance on cars in cities that is difficult for individuals to combat on their own. Our governments can even sometimes work against very innovative projects spawned by individuals. For instance, in Ontario several years ago, a service similar to the French Allo-Stop ride-sharing service—which matched riders and drivers to share gas and driving for long-distance trips—was stopped from operating by government regulations. The regulations purported to safeguard public safety and the rights of licensed for-hire drivers in the province, which does little to explain why the neighboring province of Quebec and the country of France both operate successful services on the same principle, alongside thriving bus lines with excellent public safety records.

While the planning decisions of governments may seem to be out of the control of any one person, the truth is, governments exist to serve their people. If you don't like what your government is doing, make your voice heard.

# Project 52: Attend a Public Consultation or Planning Meeting

A range of government bodies hold public consultations—your local planning department is no exception. Planning departments, as well as many other government decision-making bodies, are in many cases required by law to publicize changes and adjustments they intend to make or give permission for. Far too many of these meetings and consultations go unattended, or poorly attended.

## You will need

- A spare evening, maybe two.
- Your local newspaper. A common location for planning applications and zoning regulation changes to be posted is in the local newspaper. If you can't find them there, you might have to go digging into your local government departments, to find out when and where notices are posted.
- To have educated yourself on issues you would like addressed. For example, you may like to ask about accommodation for bicycles and walkers in the form of sidewalks and covered and secure bicycle parking. It seems commonplace lately in new developments to leave these very useful structures out entirely, in the name of cost savings. If they are in place, ask about linkages and contributions to the larger cycle and walking network in your area, or about public transport links.
- If you are attending a planning consultation, you may also want to ask questions on related issues. For example, if you are interested in increased low-

impact transportation in an existing or built-up location, remember that densification of older areas is a significant way in which we can prevent far-flung development from encouraging automobile-dependent housing. You may also want to ask about renewable energy systems, efficiency measures, waste management, water conservation, and other topics.

- The idea is not to make redevelopment difficult for planners or developers, but to ensure they have considered the range of options.

## Steps

1. Get aware of the planning process and the very real challenges of balancing competing interests that planners are trying to do. Remain aware that too often the people running these meetings have simply never considered that a bicycle rack belongs in the front of the building, closer than the cars and not further away, or that water conservation measures at an early stage of the design will save the city from upgrades years down the road.
2. Arrive hoping to make new friends, not with the intention to "tell someone off." Confrontation rarely gets anyone anywhere. For your first meeting try to find a subject that is not too close to your heart and treat it objectively. Make notes during the meeting. Raise concerns if you have them; otherwise, listen and learn. It may not be the most exciting thing in the world, but real change can come out of it.
3. Tell others about what happened, whether good or bad. Use clear language and publicize it as widely as possible. Planning issues can descend into "us-and-them" scenarios; try not to let it get there by being aware that the planners are people too, and are not necessarily anti-anything, even though it may seem otherwise.

## Stepping out

Sometimes, though everyone realizes it's a good idea, and is needed, the things that we think should happen in the world take longer than we want, or just don't seem to happen at all. In these cases, some may find it necessary to step out, just a bit.

In Toronto a major cross-town route without bike lanes resulted in the death of two cyclists in one week in 2008. Articles were written in the local newspaper about how Bloor St needed a bicycle lane, and many people opined loudly that the days of on-street parking had to finally come to an end. The city seemed to have been studying the issue since the 1970s, and though they have a fantastic plan for a great expansion of bicycle routes, it didn't seem to be getting there quickly enough for some people.

Building your own bicycle path is not something I can encourage you to do. It could be dangerous, and is quite illegal. But some fearless geniuses have done exactly that, in Toronto and around the world. The process for making your own bike paths are available on the Internet if you are interested in pursuing this. Simply use spraypaint in a more or less straight line down the edge of the roadway; if you are going for accuracy, paint a bike sign. The painters in Toronto found that people in cars thought they were real right away and immediately stayed out of them. I met Italian bike lane painters who took pride in an article in their local paper that criticized the local council for painting crooked bike lanes, when in fact the group had painted the "bike lanes" themselves using a specially designed trailer to hold the paint and an automatic trigger to start making lines!

Note that your actions could cost your municipality, which is likely to decide to remove them using funds which could have gone towards putting in actual bike lanes. But there is a certain level of frustration if a person tries the advocacy route repeatedly without success. Lines of paint are not expensive and our governments need to know that putting them in the right places is important to us.

Failing these types of direct action, smaller actions can be used to make others know that bikes have rights too. Bigger human-powered vehicles, like the ones we explored earlier, are a step in taking space for sustainable transportation. For narrower vehicles without separate direct paths to use, remember that it is your right as a user of the road to use a whole lane of traffic. Riding in the middle of the lane is a useful technique, which ensures a wide berth by motorists. Though you may not be popular with them, stand your ground. Awareness that sustainable modes of travel have a right to use a proportionate amount of space creates safer streets for all users.

# Index

## D



## E



## F



## G

## H



## I



## J



## K



## L



## M



## N



## O

## P



## Q



## R



## S